Nuclear Legacies

Lexington Studies in Political Communication

Series Editor: Robert E. Denton Jr., Virginia Tech

This series encourages focused work examining the role and function of communication in the realm of politics including campaigns and elections, media, and political institutions.

TITLES IN SERIES:

Governing Codes: Gender, Metaphor, and Political Identity, By Karrin Vasby Anderson and Kristina Horn Sheeler

Paving the Way for Madam President, By Nichola D. Gutgold

Maryland Politics and Political Communication, 1950-2005, By Theodore F. Sheckels

Images, Issues, and Attacks: Television Advertising by Incumbents and Challengers in Presidential Elections, By E. D. Dover

Democracy as Discussion: Civic Education and the American Forum Movement, By William Keith

Nuclear Legacies: Communication, Controversy, and the U. S. Nuclear Weapons Complex, Edited by Bryan C. Taylor, William J. Kinsella, Stephen P. Depoe, Maribeth S. Metzler

Picturing China in the American Press: The Visual Portrayal of Sino-American Relations in Time *Magazine, 1949–1973,* By David D. Perlmutter

Post-9/11 American Presidential Rhetoric: A Study of Protofascist

Discourse, By Colleen Elizabeth Kelley

Making a Difference: A Comparative View of the Role of the Internet in Election Politics, Edited by Stephen Ward, Diana Owen, Richard Davis, and David Taras

Seen and Heard: The Women of Television News, By Nichola D. Gutgold

Nuclear Legacies

Communication, Controversy, and the U.S. Nuclear Weapons Complex

Bryan C. Taylor,
William J. Kinsella,
Stephen P. Depoe,
and Maribeth S. Metzler

LEXINGTON BOOKS

A division of
ROWMAN & LITTLEFIELD PUBLISHERS, INC.
Lanham • Boulder • New York • Toronto • Plymouth, UK

LEXINGTON BOOKS

A division of Rowman & Littlefield Publishers, Inc.
A wholly owned subsidiary of The Rowman & Littlefield Publishing Group, Inc.
4501 Forbes Boulevard, Suite 200
Lanham, MD 20706

Estover Road
Plymouth PL6 7PY
United Kingdom

First paperback edition 2008

British Library Cataloguing in Publication Information Available

Library of Congress Cataloging-in-Publication Data
Nuclear legacies : communication, controversy, and the U.S. nuclear weapons complex / edited by Bryan C. Taylor . . . [et al.].
p. cm. — (Lexington studies in political communication)
Includes index.
1. Nulcear weapons plants—Environmental aspects—United States. 2. Nuclear weapons plants—Social aspects—United States. 3. Nuclear weapons—Moral and ethical aspects—United States. 4. Nuclear industry—Government policy—United States. 5. Environmental policy—United States—Citizen participation. I. Taylor, Bryan C.
TD195.N85N825 2007
363.738—dc22 2006035258

ISBN-13: 978-0-7391-1904-4 (cloth : alk. paper)
ISBN-10: 0-7391-1904-4 (cloth : alk. paper)
ISBN-13: 978-0-7391-1905-1 (pbk. : alk. paper)
ISBN-10: 0-7391-1905-2 (pbk. : alk. paper)

Printed in the United States of America

∞™ The paper used in this publication meets the minimum requirements of American National Standard for Information Sciences—Permanence of Paper for Printed Library Materials, ANSI/NISO Z39.48–1992.

Table of Contents

Acknowledgments vii

1 Introduction: Linking Nuclear Legacies and Communication Studies
Bryan C. Taylor and William J. Kinsella 1

Part One The Discourse of Officials and Stakeholders of Nuclear Weapons Production

2 Convergence and Divergence in the Public Dialogue on Nuclear Weapons Cleanup
Jennifer Duffield Hamilton 41

3 Becoming Hanford Downwinders: Producing Community and Challenging Discursive Containment
William J. Kinsella and Jay Mullen 73

4 Regional Communication and Sense of Place Surrounding the Waste Isolation Pilot Plant
Eric L. Morgan 109

Part Two Organizing the Past, Present, and Future of Nuclear Weapons Production

5 Cold War Triumphant: The Rhetorical Uses of History, Memory, and Heritage Preservation within the Department of

Energy's Nuclear Weapons Complex
Jason N. Krupar and Stephen P. Depoe 135

6 TRUTH Is Generated HERE: Knowledge Loss and the Production of Nuclear Confidence in the Post–Cold War Era
Laura A. McNamara 167

7 (Forever) At Work in the Fields of the Bomb: Images of Long-Term Stewardship in Post–Cold War Nuclear Discourse
Bryan C. Taylor 199

Part Three Critical Response

8 Response: Nuclear Legacies and Opportunities for Politically and Ethically Engaged Communication Scholarship
Tarla Rai Peterson 237

Index 255

About the Contributors 265

Acknowledgments

Bryan Taylor wishes to thank Michele Jackson and the Department of Communication at the University of Colorado, Boulder, for funding supporting the publication of this volume. Bill Kinsella wishes to thank the many Hanford stakeholders and officials who have helped him make sense of the conditions examined herein.

The editors wish to thank Molly Ahearn, Robert Denton, and Joseph Parry at Lexington Books for their many efforts in support of this project. We also wish to thank the office of U.S. Representative David Price (NC) for research assistance.

1

Introduction: Linking Nuclear Legacies and Communication Studies

Bryan C. Taylor and William J. Kinsella

As we finalize this introduction in late 2006, the news is filled with stories of nuclear threat. North Korea has provoked international alarm and sanctions for its announcement of an underground nuclear test. A few months prior, it attempted unsuccessfully to test-launch a missile that may someday prove capable of striking the United States. Meanwhile, international governments, led by the Bush administration, continue to urge Iran to halt its controversial uranium enrichment program, and to instead "observe its international obligations" to cease nuclear proliferation. To these pleas, threats, and proposed alternatives, the Iranian government has responded (so far) with missile tests, indignant assertions of an inalienable right to develop nuclear technologies, and plans to construct a new heavy water production facility.[1]

As communication scholars interested in nuclear issues, we react to these stories with keen ambivalence. We are American citizens who grew up during the Cold War, and so we recognize in them a familiar call to alarm concerning *other nations'* nuclear weapons programs. As a form of rhetoric, this call evokes vague feelings of dread and resignation—the affective residues of an earlier era marked by "duck and cover" drills, allegations of "bomber" and "missile gaps," and official antagonism toward "the Evil Empire." It provokes the anxiety and vigilance of American audiences, and directs them outward to shadowy images of exotic and dangerous entities, including both "rogue states" and the nuclear "terrorists" they may facilitate. Unable to fully locate, understand or control those entities, however, ordinary citizens must rely (at least in part) on military and political officials to interpret these conditions, and to respond to them adequately. As a result, for those citizens, the nuclear threat persists as an international spectacle that threatens to breach its staging, and to engulf its captive audience.

In a September 2, 2006 editorial, *The New York Times* described the growing risk from nuclear proliferation as an ironic "nuclear legacy" of the Bush administration's unilateral and militarist foreign policy. We do not dispute the enduring, complex, and dangerous qualities of international proliferation. We argue in this volume, however, that this danger is not the only legacy of nuclear weapons production in the post–Cold War era. In important ways, all aspects of the global nuclear apparatus are connected. Radioactive fallout from decades of atmospheric testing by nuclear states, for example, has spread worldwide. Fissile materials such as processed uranium and plutonium circulate through an international network of military production reactors, waste treatment facilities, and civilian power reactors.[2] Although these images suggest the expansive scope of nuclear matters, this volume deliberately narrows that focus. It allows readers to enter (or perhaps to revisit) the *domestic* scene of America's Cold War–era nuclear weapons production, and to consider the communication phenomena that have developed there following the declared "end" of the Cold War.

We acknowledge that this focus is partial. We justify it, however, by citing the benefits of careful attention to a scene that continues to be neglected and distorted in the consciousness of many Americans. In that scene, we find a vast, dispersed network of laboratories, factories, test sites, cleanup sites, waste repositories, and local communities known as "the nuclear weapons production complex."[3] There, we find a story that is more difficult for Americans to assimilate. This is because that story combines sharply opposing themes of continuity and change, decay and renewal, innocence and guilt, trust and betrayal, openness and secrecy, and citizen activism and government power. Accordingly, our tolerance for ambiguity and contradiction is tested as we consider the recent evolution of America's own nuclear weapons program. That evolution, we argue, may be gauged by considering changing relationships between the knowledge, institutions, identities, policies, labor, technologies, and discourses that constitute the current nuclear weapons complex. This volume pursues *that* story, and we believe it offers significant opportunities for learning. Through this story, we learn about the legacies of Cold War nuclear weapons production for our nation's environment, public health, and worker safety. We learn how these legacies shape the ongoing efforts of our civilization to understand and influence the development of nuclear weapons.

Uniquely then, we consider nuclear legacies to include not only strictly *material* phenomena associated with weapons and waste, but also *practices and processes of communication*. For example, Kinsella (2005a) has analyzed how, in informing and influencing audiences, nuclear speakers[4] have historically relied on discursive tropes such as *mystery*, *potency*, *secrecy*, and an *entelechial* admonition to bring scientific and technological possibilities to their fruition. In this volume, we are concerned with how these kinds of *communication legacies* have shaped—and have been shaped by—the interaction, collaboration, and deliberations of nuclear officials and stakeholders in the post–Cold War era. As a result, our focus is on how communication constitutes and transforms the significance of various objects in their shared world (e.g., institutional reputations, natural environments, knowledge claims, etc.). How, for example, do speakers symbolically imbue these objects with qualities such as *necessity*, *inevitability*, *legitimacy*, *priority*, *authority*, and *validity*?

In various ways, the chapters collected in this volume depict how speakers associated with the post–Cold War nuclear weapons production complex draw upon various discourses (e.g., of risk analysis) that are available to them as a result of their positioning in social worlds (e.g., as professional scientists). The chapters demonstrate how these speakers perform these discourses through routine practices such as asking and responding to questions in public hearings, and interpreting the data from simulated nuclear explosions. They show how these performances serve to formulate the interests of nuclear groups, and to represent those interests in both official forums of decision-making and unofficial scenes of nuclear-cultural life. They reveal how, in using discourse to achieve their goals, speakers are enabled and constrained by its logical structures, and by the ideologies that animate it with preferred images of what is true, real, and good (and their opposites) about nuclear weapons in the post–Cold War era.

As a result, the communication phenomena depicted in these chapters indicate a larger cultural and historical struggle over the meaning and consequence of nuclear weapons production in post–Cold War America. They suggest how communication develops ongoing relations of collaboration, competition, and conflict among affected nuclear groups. We learn how communication produces affiliation and difference, both within and between the members of these groups. We learn how their divisions and conflicts may be managed communicatively, and what the consequences of those efforts may be. As a result, our understanding of the granularity, integrity, and consequentiality of social action in the post–Cold War nuclear weapons complex is enhanced. We see how, as it is realized through communication, the nuclear *word* contributes to the post–Cold War nuclear *world* (see Dowling 1987, 208). We are better able to imagine and practice new ways of speaking about nuclear weapons development—ways that may advance the interests of dialogue, democracy, and justice.

NUCLEAR WEAPONS AND COMMUNICATION STUDIES: A BRIEF OVERVIEW

This volume extends a program of research within the communication discipline that is generally concerned with understanding, evaluating, and transforming the nuclear condition. Taylor (1998a) has recounted the emergence of this research program during the 1980s: That account depicts a convergence of late Cold War anxieties, rising antinuclear social movements, and an interdisciplinary project known as "nuclear criticism" which encouraged communication scholars across different subfields to examine the symbolic dimensions of nuclear weapons phenomena. This body of work reflects four traditions utilized by those scholars to theorize and critique the relationship between nuclear weapons and communication.

First is a *dramatistic* tradition, largely derived from the work of Kenneth Burke and Erving Goffman (see Bjork 1992; Cheney 1991; Duffield 2003; Fisher 1984; Hubbard 1998; Kauffman 1989; Kinsella 2005a; Meyer 1995; Schiappa 1989; Tietge 2002; Toker 2002; von Meier, Miller and Keller 1998; Williams 1988; see also Benford 1993). This tradition depicts humans as uniquely symbol-using creatures who are continuously engaged in making nuclear weapons a *meaningful* condition of their social reality. They satisfy this need through rhetorical practices such as *naming*, *defining*, *framing*, *identifying*, and *narrating*. These practices are simultaneously stylized, constitutive, and strategic. As elements of nuclear rhetoric, they shape the attitudes and actions of audiences towards nuclear phenomena, inducing their acceptance of particular stories that explain and interpret the nuclear condition. These practices establish the identities of nuclear actors (e.g., as victims and victimizers), attribute and moralize motives for their actions (e.g., arrogance and naiveté), configure the relationships between actors (e.g., as a criminal conspiracy; see Kauzlarich and Kramer 1998), and recommend strategies for restoring breaches of social order. Through these practices, speakers perform essential, overlapping social functions such as establishing credibility with their audiences, overcoming division (e.g., through the depiction of similar interests), mobilizing support for policy proposals from diverse constituencies, challenging authority, and defending their own authority against counter-challenges.

Unfortunately, these practices may also exacerbate an innate human compulsion to fulfill mythical narratives that prescribe desirable—and seemingly inevitable—end-states of nuclear development (i.e., *teleologies*). This tendency is fueled by the abstract quality of symbols, and is most clearly manifest in the continuous "perfection" by nation-states of nuclear weapons as technologies of foreign policy and military strategy (Brummett 1989). Here, scholars emphasize the role of language in creating *frames* (or, for Burke, "terministic screens") that serve group interests by selectively emphasizing and interpreting features of the nuclear world, while deflecting alternatives that are inconvenient and contradictory to those interests. Due to the high-stakes controversies surrounding nuclear weapons development, most—if not all—of its associated scenes are characterized by competition between multiple frames and narratives. Of particular interest to communication scholars are the ways in which military, scientific, and government elites use nuclear rhetoric (e.g., technical rationality) to accomplish politically regressive outcomes. These outcomes include: mystifying audiences; obscuring the contingency, historicity, and human agency underlying policies and operations; promoting hierarchical and authoritarian institutions (such as nuclear "priesthoods"); and neutralizing dissent. Those scholars also note how opposing groups challenge dominant nuclear frames by asserting competing logics and vocabularies (e.g., of the *comic* or *ironic*)—and how these efforts elicit rhetorical counterinsurgency in kind.

A second tradition of nuclear communication research involves *modernist critical theory*, largely influenced by the work of Frankfurt School founders Theodor Adorno and Max Horkheimer, and their successors (see Katz and Miller 1996; Metzler 1998; Ratliff 1998; Taylor 1990; see also Dayton 2002; Futrell, 2003; Glass 1993; Hardert et al. 1989; Kinney and Leschine 2002; Kuletz 1998; Mehta 1998; Skillington 1997). This tradition is primarily concerned with the quality of citizen participation and deliberation surrounding official nuclear policies. This work variously orients to Jurgen Habermas' controversial concepts of the "critical public sphere" and "communicative action" as sources of normative ideals for evaluating actual communication by participants in the development and enactment of these policies (e.g., in required public hearings for environmental impact statements). These ideals privilege the integrity of deliberative process over the achievement of a particular outcome. They generate a number of prescriptions, including: that deliberation of consequential nuclear matters should be accessible to public participation; that citizens should be sufficiently motivated and informed to engage those matters; that speakers should have access to—and competency in—the available means of influencing outcomes (e.g., placing items on discussion agendas and setting discussion rules); that participants should debate issues openly, democratically, and rationally, deferring their pre-existing differences of status and expertise; that participants should be able to openly reflect on the intelligibility, truthfulness, sincerity, and appropriateness of offered claims; that participants should seek to reach consensus through the use of practical reasoning concerned with the quality of their shared lifeworld; and finally, that participants' deliberation should lead not only to the formation of public opinion, but also actually influence official decision-making and associated operations.

Here, communication scholars focus on historical conditions and institutional practices that inhibit the attainment of this ideal, and that perpetuate the structures and cultures of nuclear domination. Frequently cited are the wartime development of nuclear weapons in a climate of urgency and secrecy; the subsequent postwar embrace of those weapons as essential instruments of national security; the exemption of the expanding Cold War–era weapons production complex from traditional regulatory oversight; the tendency of nuclear officials and corporate contractors to suppress and distort information concerning the risks of operations posed to worker safety, public health, and the environment (e.g., through the use of assertions, jargon, and euphemism); and a cynical approach to public participation by nuclear officials concerned with "engineering" citizen compliance with preferred policies (e.g., through programs of "education" that exclude from consideration citizen values concerning nuclear risk; Taylor 2003b). Here, the focus is on practices of power (e.g., arbitrarily constraining the scope of discussion) that alternately enhance or preclude the possibility of mutual, authentic, and transformational dialogue between nuclear officials, citizens, and the organized members of nuclear "counter-publics" (e.g., social movements such as the Nuclear Freeze). Here, the development of nuclear weapons forms a rich opportunity to consider evolving relationships between the instrumental rationality dominating modern society (i.e., that narrowly privileges profit, efficiency, and productivity), and the organic, vernacular discourses of the social lifeworld.

The tone of these studies ranges from tragic depiction of the successful containment of citizen discourse by officials to dramatic exploration of the creative, tactical moves made by counterpublics to breach and revise those structures (such as redefining the official criteria by which topics and concerns may be considered "relevant"). Here, drawing on the work of communication theorists such as Stanley Deetz (1991, 1995), organizational communication scholars play an important role in assessing the "corporate colonization" of public deliberation and the prospects of genuinely democratic communication—both within nuclear organizations and between their members and stakeholders (stemming, for example, from robust conceptualizations of *forum* and *voice*). Further work in this tradition, however, must account for growing evidence that normative criteria may not be inherently necessary or useful for producing outcomes deemed satisfactory by participants (Lawless, Bergman and Feltovich 2005; Santos and Chess 2003).

Third is a closely associated tradition of *argumentation theory and analysis*. This tradition extends critical-theoretical concern with the integrity of public deliberation by augmenting its account of the relationships between spheres and "fields" of discourse. It does so by drawing on the work of argumentation theorists such as Stephen Toulmin and Chaim Perelman concerning the unique logical conventions that govern the articulation of claims, evidence and warrants in institutional communication (e.g., religion, science, and government; Goodnight 1982). Here, argumentation scholars untangle the exceedingly complex and paradoxical conventions of nuclear policy discourse (e.g., surrounding the varieties of nuclear deterrence; Dauber 1993). In their study of "argument spheres," critics examine how nuclear institutions regulate the discourse formally falling under their jurisdiction, how they ambitiously (and often coercively) impose their argument conventions on other spheres, how they strategically appropriate competing conventions to establish their own authority and legitimacy, and how they maintain those attributes by undermining opponents and neutralizing dissent.

Communication scholars working in this tradition focus on the consequence of this activity for public deliberation of nuclear issues. A frequent theme involves the corrosive effect produced by unreflective or authoritarian imposition of conventions derived from technical, specialized discourse spheres (e.g., of scientific expertise) (Farrell and Goodnight 1981; King and Petress 1990; Prosise 1998). Schiappa (1989), for example, has analyzed two dialects of official "nukespeak" which help to normalize the nuclear condition: *domestication*, which renders nuclear topics palatable by associating them with personal, familiar, and innocuous images (see also Cohn 1987; Taylor 1997a), and *bureaucratization*, which sanitizes and abstracts nuclear issues, rendering them either neutral or inaccessible to lay debate. In both cases, Schiappa concludes, "It is clear that *with regard to nuclear issues, the public has been conceived as a crowd to be calmed rather than co-creators of public policy* . . ." (Schiappa 1989, 260; emphasis in original).

Argumentation scholars have also advanced this tradition by noting that some specialized spheres of nuclear argument (such as arms control) are conflicted, interdisciplinary ecologies of discourse. As such, they do not yield simple identification of consistent generic conventions. Instead, argument scholars must use critical and interpretive methods to untangle and evaluate the arcane and competing conventions used by speakers arguing in these spheres (Dauber 1988; Hynes 1988). Finally, it is worth noting that five of the six "questions of theoretical interest" posed by a prominent argumentation scholar at the close of the Cold War still seem highly relevant for our current nuclear moment: "The hardening of historical interpretation, the fragmentation of public discourse . . . the strategic options of counter-movements, the search for alternative sources of reason, and the power of critique to undermine the legitimacy of a specialized language" (Goodnight 1988, 142).

Lastly, we identify a tradition of nuclear communication scholarship characterized by a postformalist mélange of *semiotic*, *poststructuralist,* and *dialogic* theories. Generally, these theories are united in their challenge to realist and cybernetic paradigms of nuclear communication that depict it as a transparent medium for reflecting pre-existing, objective nuclear truth, and for mechanically transmitting messages (or, "communications") between sources and audiences that encode nuclear truth as "information." Instead, these theories emphasize the role of discourse in *constituting* knowledge of the nuclear condition by employing familiar conventions of representation whose effectiveness lies in their ability to both depict material phenomena *and* sustain cultural hegemony. Here, the underlying assumption is that officials "manufacture" public consent to the nuclear condition through the pervasive circulation of discourses that structure the possibilities for meaningfully existing and acting within nuclear institutions. One example is the recurring claim that, since nuclear weapons are irreversible and necessary for national security (i.e., "You can't put the genie back in the bottle"), protest is irrational and illegitimate.

One stream of theory here emphasizes deconstruction of the inevitable contingency, prematurely closed structures, and ideological productivity of nuclear language. In this view, despite their pretense to coherence, rationality, and neutrality, official nuclear discourses are typically dependent on the use of both figurative elements such as metaphor (e.g., of "containment"; Kinsella 2001), and on "external" discourses such as myth, religion and literature to sustain their authority and legitimacy (Taylor 2001). Nuclear critics subsequently intervene in these discourses by clarifying their hybrid, over-determined (and often schizophrenic) status, thus opening up new possibilities for their ordering of nuclear understanding and action (Williams 1988; see also Chaloupka 1992; Schwenger 1992). This tradition relies on Jacques Derrida's (1984) landmark essay establishing the tortured ontology of nuclear weapons as—mostly, so far, thankfully—an object of "fabulously textual" simulation and projection. In that essay, Derrida noted the ominous and perverse condition surrounding nuclear discourse: if its ultimate, apocalyptic referent of global nuclear war were ever realized, that event would be both unrepresentable in language, and would destroy the material and the social worlds which establish language as a relevant resource for human living. However, as several critics have noted (Norris 1994; Ruthven 1993; Taylor 1998a), Derrida's provocative formulation ignores key historical instances of nuclear materiality, including the U.S. bombings of Hiroshima and Nagasaki and—most relevant for this volume—the devastating consequences of nuclear weapons development for public health, worker safety, and the environment (Makhijani, Hu, and Yih 1995).

This revision of the relationship between nuclear symbolism and materiality also challenges conventional political wisdom that nuclear weapons have not been "used" (i.e., in military combat) since the American bombings of Japan. Here, we must note the restrictive application of the term "use" in this claim. Alternately, we may consider the meaning of this term in its most basic sense: as the human employment of technology to achieve goals in a manner that produces material effects. If we apply this definition of "use" to the case of nuclear weapons development, we are suddenly confronted with a long list of witnesses—including nuclear workers, community members and environmental activists. These voices clamor to testify that nuclear weapons have indeed been "used," and have produced effects such as radiation-related illness that have not been "useful" to the life-forms they have marked. As a result, we are invited to question how, when, and by whom nuclear weapons may be "used"—and may be deemed "useful."

As a primarily humanistic product of literary theory, linguistics, and philosophy, nuclear criticism has typically directed its attention to cases of literature, popular culture, and political discourse. As a result, it has often failed to engage the more mundane—but no less important—organizational and industrial dimensions of the nuclear apparatus. Glimmers of recognized need for this work occasionally surface in humanistic scholarship. One example is the following question, posed by Kalaidjian (1999) in his critique of existing conceptualizations of the nuclear-cultural audience: "Must we not," he asks, "draw ethicopolitical distinctions among perpetrators, bystanders and victims of nuclear *production* as it is imbricated with nuclear annihilation?" (318; emphasis in original).

As members of a discipline that integrates humanistic and social scientific traditions, nuclear communication scholars are poised to fill this gap. They have done so partly by drawing on the work of Michel Foucault to explore the relations between power, knowledge, and discourse operating in nuclear-institutional settings. Specifically, Foucault's complex *oeuvre* offers resources for conceptualizing how those institutions regulate communication by establishing "who can speak with authority about nuclear topics, what can and cannot be said about them, and in what settings this discourse can take place" (Kinsella 2005a, 52). Under these conditions, it is as accurate to say that institutional discourses *speak their speakers* as the other way around. Over time, speakers produce "statements" that characterize distinctive knowledge/power regimes within institutions, and that contend for influence with other statements associated with competing regimes operating throughout their environments. The configurations of these statements may subsequently be read as "formations" of nuclear discourse defining particular historical periods such as "the post–Cold War."

Foucault's work also enables communication scholars to analyze the "disciplinary effects of power upon individuals who are incorporated into institutional nuclear systems" (Kinsella 2005a, 63; see also Kinsella 1999, 2005b). This focus includes the interlinked situations of nuclear professionals and citizens, whose potential for subjectivity and agency in the reproduction of nuclear hegemony is both intensively cultivated (e.g., as loyalty, patriotism, and productivity) and prohibited (e.g., as dissent and treason) by powerful institutions obsessed with security and secrecy. As a result, we are better able to appreciate the episodes and practices that characterize these contexts (e.g., job interviews, whistle-blowing, etc.) that might otherwise be glossed or ignored in humanistic conceptions of the nuclear condition (Taylor 1992, 2002). These studies depict the fluid and productive circulation of power in the micro-practices of everyday nuclear-institutional life. They show how this circulation produces both mechanisms of conformity, and also opportunities for resisting and disrupting dominant regimes (e.g., through subjects' identification with professional integrity over immediate organizational loyalty in their evaluation of controversial nuclear-scientific data; see also Gusterson 1996).

A final strand in this tradition involves a *dialogic* approach to nuclear culture, best exemplified in the literary and cultural theory of Mikhail Bakhtin. This strand of theory emphasizes a politics of representation involving the relational (e.g., "intertextual") properties of discourses and "utterances" made in the context of evolving nuclear-cultural controversy (Benson and Anderson 1990; Mechling and Mechling 1995; Taylor 1993a, 1993b, 1996, 1997a, 1997b, 1997c, 1998b; see also Mehan, Nathanson and Skelly 1990; Mehan and Wills 1988; Wertsch 1987). In this view, nuclear culture—and associated scenes such as the nuclear weapons production complex—are "noisy" sites, "aswarm" with the multiple and conflicting voices of nuclear interests.[5] A *partial* list of these interests includes: pacifists, environmentalists, scientists, arms-control negotiators, federal and state regulators, military officials and veterans, corporate contractors, legislators, artists and performers, historians, feminists, and the local governments and community residents surrounding nuclear weapons production, testing, and cleanup sites. In this polarized and agitated climate, voices simultaneously affiliate with, diverge from, interanimate, and transform each other. No nuclear utterance, in this view, is wholly original and there is no assumption that communication will necessarily or dialectically progress toward Truth. Instead, nuclear-cultural conversations create moments and spaces in which American citizens may interpret and express the evolving reality of nuclear weapons development. As a result, the nuclear-critical task is to track the structure and process of this vital, ongoing exchange (which occurs both within and across particular texts and contexts). The goal is not simply to inventory nuclear voices and utterances, but also to consider how their relationships develop, both within and across institutional sites and discursive genres. Of particular interest here are dialogic displays of evocation, provocation, allusion, quoting, parody, attack, defensiveness, and reciprocity. All of these are variants of the general qualities of anticipation and responsiveness which orient a nuclear utterance in time, both retrospectively toward history, and prospectively toward the future. In light of Derrida's poststructuralist argument concerning the ontology of nuclear discourse, there is an implicit ethic to this critical position: As long as the nuclear conversation continues, the very worst outcome is—at least for now—averted (see Taylor 1998a, 306–307).

This overview can only characterize these traditions of nuclear communication scholarship. It cannot fully elaborate their internal diversity[6], their overlapping ancestries,[7] their derivatives,[8] their minor and neglected variants,[9] or the ways in which they are creatively combined to accomplish situated projects.[10] Nonetheless, this body of work has established commitments that distinguish it from related work conducted in other disciplines such as anthropology, political science, and sociology.[11] Those commitments may be summarized as the recognition that nuclear phenomena can only be understood and engaged through simultaneously partial and constitutive discourses that are suffused with historical and cultural relations of power. As a result, nuclear communication is not merely a neutral or transparent means of transmitting preconstituted nuclear information (e.g., as "facts"; Bazerman 2001). Instead, it is a set of symbolic practices that are alternately creative, courageous, routinized, paradoxical, oppressive, contested and subversive. These practices, conducted with and for others, *make the nuclear world meaningful, and capable of being acted toward*. Ideally, they engage others ethically and accountably in deliberating questions of how we should live together within the tense and dangerous convergence of state power and high-technology (Winner 1980). As the chapters in this volume reveal, that convergence has produced for the United States both *an internationally dispersed, world-threatening nuclear arsenal* and *a domestic industrial and cultural life-world* devoted to the development of that arsenal and to the management of its legacies. It is this second object that is the focus of this volume, and we turn now to discuss its history and current status.

THE POST–COLD WAR AND THE U.S. NUCLEAR WEAPONS PRODUCTION COMPLEX AS SITES OF COMMUNICATION AND CONTROVERSY

The studies collected in this volume extend these scholarly traditions by focusing on the nuclear weapons production complex as a site of recent and continuing controversy (see Taylor, Kinsella, Depoe and Metzler 2005). Within the global nuclear apparatus, there are many specific sites from which to choose. Because these sites are all in some way connected, it is necessary to situate any particular choice in the international developments surrounding nuclear weapons since the end of the Cold War. Thus, in order to establish the broad significance of the events discussed in this volume, we begin by briefly reviewing these developments. We focus specifically on trends that may be characterized, alternately, as politically progressive and regressive (Allison 2004; Beckman, et al. 1992, 261–285; Broad 2003; Caldicott 2002; Gabel 2004/2005; Kalinowski 2004; Keller 2002; Mitchell 2000; Nolan 1999; Schwartz 2001; Taylor 2003a, 10–14).

Post–Cold War nuclear trends that are generally viewed as progressive include the following:

- The reduction (although not elimination) of the risk of full-scale, global nuclear war created by the collapse of the Soviet Union, and the openings that were created for both affirming the "military uselessness" of nuclear weapons (Rhodes 1994, 30) and for negotiating nuclear arms reduction treaties. These treaties—combined with unilateral action by nuclear weapons states—have led to a considerable decline in the global Cold War nuclear arsenal.
- The relinquishment and destruction by some nation-states (e.g., the Ukraine) of their nuclear weapon stores, the voluntary abandonment of weapons development programs (e.g., by South Africa), and the effects on some emerging programs (e.g., pre–2003 Iraq; Libya) of a combination of international diplomacy, inspections, and military interventions.
- Collaboration between the United States and Russia to secure each nation's stockpiles of nuclear warheads and fissile materials against the various threats of unauthorized diversion and use.
- A general—even if only temporary—legitimation crisis faced by U.S. defense and nuclear institutions as they were challenged to justify their Cold War–era missions and budgets in the face of a rapidly changing geopolitical environment. This crisis initially encouraged speculation concerning a "peace dividend" of reductions in both military forces and defense expenditures.
- Repeated extensions and affirmations by its signatories of the Nuclear Non-Proliferation Treaty (NPT), which provides incentives for non-nuclear nations to not develop nuclear weapons, and obligates the existing nuclear powers to pursue nuclear disarmament.
- The signing by the United States and Russia in 1992 of a comprehensive test ban treaty (CTBT) restricting the explosive testing of nuclear devices, and continued U.S. willingness to abide by that treaty despite the absence of its formal ratification.

Unfortunately, these accomplishments were laced with qualifications and complications, and were offset by trends indicating the effective adaptation by nuclear institutions to challenges posed by the post–Cold War condition. These include:

- A failure by nuclear states to continuously negotiate arms-reduction treaties that are binding, verifiable, and that yield actual dismantlement of nuclear arsenals.[12]
- The formidable inertia of the nuclear-military bureaucracies, and their cumulative success in frustrating attempts at innovation and oversight.
- The ascent of neoconservative narratives sustaining nuclear weapons as a viable means of achieving U.S. national security. In these narratives, nuclear weapons are depicted as requirements for performing both continuing and new missions, including: sustained deterrence of existing nuclear states such as China; unfolding deterrence of new and "rogue" nuclear states such as North Korea; and deterrence of enemy use of biological and chemical weapons of mass destruction. Following the end of the Cold War, these narratives were developed and encoded in official policy and planning documents such as Presidential Decision Directives, Nuclear Posture Reviews, and nuclear-war-fighting guidance. They have subsequently fueled resistance to the ratification and continued observation of arms control treaties (e.g., the U.S. Senate's rejection of the CTBT in 1999), and have helped to re-inflate declining federal defense budgets. Encoded in defense policy, these narratives have also directed the military to update its strategies for using nuclear weapons against prospective nuclear states (e.g., Iran), and have spurred consideration of the development of smaller-yield nuclear weapons (such as "bunker-busters") deemed necessary for use in post–Cold War battle environments.
- The continued promotion and development of military defenses against "new threats" of accidental and strategic missile launch, leading the United States to withdraw in 2002 from the 1972 Anti-Ballistic Missile Treaty, and to persist in a costly and technologically dubious project that exacerbates nuclear risk by increasing the viability of a U.S. first strike against other nations.
- Growing nuclear risk created by the horizontal proliferation of nuclear weapons programs among states such as North Korea and (potentially) Iran, and by the sobering possibility of nuclear terrorism resulting from the unauthorized acquisition or construction of a nuclear weapon, or the construction of a radiological dispersal device (or "dirty bomb"). It has been widely observed that the exceptionalism and unilateralism currently dominating U.S. foreign policy—combined with failure to adequately fund efforts to secure global nuclear weapons arsenals and stores of

weapons-grade nuclear materials—undermine both existing programs and proposed initiatives intended to reduce this risk.

These conditions have all—for good reason—received significant attention and generated urgent policy debate. For many Americans, however, they also form a thick canopy which deflects their awareness away from recent developments in the domestic nuclear weapons production complex, and from the consequences of those developments for these conditions.[13] We now turn to the history and status of that site, emphasizing its relevance for communication scholarship.

THE U.S. NUCLEAR WEAPONS PRODUCTION COMPLEX: HISTORY, MATERIALITY, AND DISCOURSE

Energy and environmental scholar Arjun Makhijani (2006) identifies the origin of the U.S. nuclear weapons program with the specific date of October 9, 1941. On that date, responding to an accumulation of scientific and political developments, President Franklin Roosevelt directed Vannevar Bush, head of the government's Office of Scientific Research and Development, to undertake an evaluation of the feasibility and cost of developing a nuclear weapon. Although all such historical punctuations are ultimately artificial, that date represents an important merger of scientific and institutional power; Makhijani argues that once that merger was accomplished, a crucial course was set. By the end of 1942, three key developments were complete: nuclear research and development had been consolidated as a military project under the direction of General Leslie Groves, Enrico Fermi's scientific team had produced a self-sustaining chain reaction at the University of Chicago, and President Roosevelt had approved a full-scale program of facility construction and weapons production. Although it was already understood by its founders to be an effort of unprecedented scope and complexity, over subsequent decades that program would expand far beyond their original, ambitious vision.

Initially, facilities in Clinton (now Oak Ridge), Tennessee; Hanford (now Richland), Washington; and Los Alamos, New Mexico, housed the project's primary research and production activities.[14] Beginning in the 1950s, however, Cold War imperatives led to the creation of major new government-owned, contractor-operated facilities at sites including Fernald, Ohio; Rocky Flats, Colorado; and Savannah River, South Carolina. In addition to the original research facility at Los Alamos, new laboratories were established at Albuquerque, New Mexico (Sandia Laboratories); and Livermore, California. Ultimately, the complex would encompass seventeen primary sites occupying

3,900 square miles in thirteen states, as well as approximately 300 smaller sites contributing to an elaborate, dispersed network of activities including research and development, industrial production, and testing. Over time, the administration of this evolving system would pass from the original Manhattan Project, to the U.S. Atomic Energy Commission, to the short-lived Energy Research and Development Administration, and finally, to the U.S. Department of Energy (hereafter, DOE). By 1992, this system had employed approximately 650,000 people, spent approximately $370 billion, and produced more than 70,000 nuclear weapons (Schwartz 1998; USDOE 1995, 1997; Zuckerbrod 2001). Simple arithmetic illustrates the scale of this industrial enterprise: over a forty-five-year period the system mass-produced an average of more than 1,500 weapons per year, or more than four weapons per calendar day.

A series of steps commonly known as the nuclear fuel cycle has provided the basis for these production activities, encompassing uranium mining, milling, and enrichment; nuclear reactor fuel fabrication; fabrication of "targets" for irradiation to produce tritium (the radioactive form of hydrogen used in thermonuclear weapons); reactor operations; chemical reprocessing of irradiated fuel to extract plutonium and other fissile materials; and storage and/or disposal of the massive amounts of nuclear and chemical waste that accompany these processes. The resulting weapons-grade materials were then converted into forms suitable for assembly into weapons, and final assembly was accomplished at the Pantex facility near Amarillo, Texas. Pantex is now the site for a post–Cold War program to disassemble decommissioned weapons, but it is said that the disassembly proceeds more slowly than did the production work, and with a great deal of caution, because the weapons were not designed to be taken apart. Ironically, it appears that the creators of these devices did not anticipate a scenario in which they would be peacefully deconstructed.

A similar lack of foresight attends the environmental consequences of decades of nuclear weapons production. The various steps in the production process have resulted in a broad range of hazards for nuclear workers, local and regional host communities, and the public at large (see Makhijani, Hu and Yih 2000; USDOE 1997). Those hazards include airborne releases of radioactivity and toxic chemicals; contaminated soil, groundwater, and surface water; solid, liquid, and gaseous wastes, often stored under severely compromised conditions; and an accumulation of radiological and fissile materials potentially susceptible to appropriation by terrorists or aspiring nuclear proliferators. The implications of policy decisions regarding storage, disposal, and transportation of these materials are expansive and technically complex, and have become the focus of deliberation, negotiation, and contention in a broad range of settings.

Following the increasing revelation of failures and scandals associated with this history of operations, and partly in keeping with the apparent end of the Cold War, the DOE temporarily ceased weapons production operations in the early 1990s. During this period, the DOE created an Office of Environmental Management (EM) to oversee cleanup and remediation efforts at the various sites in the weapons complex and to develop two repositories for permanent disposal of nuclear wastes. These repositories are the Waste Isolation Pilot Plant (WIPP) in New Mexico, opened in 1999 and restricted to storage of "transuranic" nuclear waste, and the long-planned but politically-contested (and thus still-uncertain) Yucca Mountain site in Nevada, developed for disposal of irradiated nuclear fuel and high-level radioactive waste. Also during this period, publicity-weary and litigation-chastened DOE officials formalized a temporary commitment to policies of "openness" concerning the legacies of Cold War–era weapons production. As a result, EM program staff not only communicated with the public in traditional contexts such as environmental impact studies and public hearings, but also began to operate in new and different contexts. These included the negotiation of consent agreements formalizing regulatory relationships among the DOE, the U.S. Environmental Protection Agency (EPA), and state environmental agencies, as well as the development of site-specific advisory boards (SSABs) comprising local and regional stakeholder groups.

The communication associated with these sites represented unprecedented opportunities for nuclear officials and stakeholders (including current and former site workers, local community government and residents, Native American tribes, and environmental and peace activist groups) to engage each other in controversial matters. In this process, the members of these groups worked intensively to gauge the possibilities and constraints characterizing this new climate, to formulate their interests and goals, and to develop and practice successful communication strategies. Through their interaction, significant new meanings and narratives about the identities and interests of affected groups, and about the history, present, and future of U.S. nuclear weapons production, were produced and contested (Taylor et al. 2005). One example involves the DOE's rhetoric in its major publications during this period, which depicted Cold War nuclear weapons production as an imminent "closed circle." Through this image, the agency implicitly asserted that its tattered reputation could be restored by successfully "cleaning up" the environmental legacies of those operations. In using this image, however, DOE officials also simplified, minimized, deferred—and generally "contained"—the volatile complexities, controversies, and incongruities that animated actual policy development and operations during this period (see Hubbard 1997; Kinsella 2001).

One example of these incongruities involves the DOE's post–Cold War efforts to maintain within its three weapon laboratories a viable capability for designing and developing nuclear weapons, and for assuring the reliability of existing weapons despite the prevailing ban on explosive nuclear testing. Following a series of security-related incidents, since 2000 these activities have been relocated under a new, semi-autonomous organizational unit, the National Nuclear Security Administration (NNSA). That institutional arrangement has further reified a well-established—but nonetheless discursively accomplished—distinction between the "weapons development" and "environmental remediation" aspects of the DOE's portfolio. Although the department's policies and rhetorical strategies since the end of the Cold War have established expectations for some (albeit, diminishing) degree of public participation in the EM program, those expectations do not extend to the weapons development programs of the NNSA. This condition has been further exacerbated in a post-2001 climate characterized by terrorism-related concerns, the consolidation of decision-making under the rubric of wartime necessity, and proposals for the development of *new* nuclear weapons such as the "reliable replacement warhead" (USDOE/NNSA 2006).

THE LEGACY TROPE

This brief overview of the nuclear weapons complex motivates our appropriation and redeployment of a concept that appears regularly in official, scholarly, and public discourse: the concept of nuclear "legacies." The title we have chosen for this volume is by no means original; indeed, the term "legacy" has been so widely applied to nuclear matters that it approaches the status of a cliché. For the physicist and nuclear enthusiast Edward Teller (1962), the "legacy of Hiroshima" was a tragic one, not only (and perhaps for Teller not primarily) because of the devastation visited upon a city, but because of the resulting, indelible association created in that event between nuclear energy and destructive violence. In Teller's view, that association impeded the subsequent development of a technology essential for an expanding industrial civilization, and stigmatized the human creativity manifested in grand accomplishments of science. Writing of "Hiroshima and its legacies," the historian Martin Sherwin (2003) focuses, instead, on issues of international affairs and the replacement of "a world destroyed" by a new world order.

Nuclear strategy theorist Thomas Schelling (2006) identifies the legacy of Hiroshima with a continuing international taboo against the military use of nuclear weapons (thus disregarding, as noted above, alternate understandings of their effectiveness in a purely deterrent role). For Schelling, "these weapons are unique, and a large part of their uniqueness derives from their being perceived as unique . . . It is simply a convention that nuclear weapons are different" (313). That convention has cultivated a "nuclear inhibition" (321) or "antinuclear instinct" (323) extending even to "small" or "tactical" nuclear weapons—although here we must note that the development and possible uses of such weapons are once again topics of public debate. A related aversion has also precluded further use of so-called peaceful nuclear explosions (318–319) that Teller and others promoted for natural gas mining, excavations of harbors and canals, and other expansive visions of the "great art of geographic engineering" (Teller, quoted in O'Neill 1998, 184).

Roff (1995) expands the legacy trope to include Nagasaki, often neglected in the shadow of Hiroshima's historical primacy, and shifts the trope's focus to the human health effects of nuclear weapons use and testing. For Roff, illness, death, and controversial medical science are the principal legacies of nuclear energy. Biologist Zhores Medvedev (1990) considers the related "legacy of Chernobyl" in terms of the implications of that 1986 Soviet nuclear reactor disaster for human health and safety, technology development, and energy policy. In an effort to promote cross-national nuclear identification, McQuerry (2000) provides a collection of student essays exploring the medical, social, and cultural legacies for children of two "atomic cities": the Ukrainian city of Slavutych, built to house residents displaced by the Chernobyl disaster, and Richland, Washington, built to house workers at the Hanford reservation, the primary site for plutonium production for U.S. nuclear weapons.

These atomic cities, deployed as atomic synecdoches, enact distinct but related interpretations of the "nuclear legacy" trope. To this inventory, Saleska (1989) adds the related legacy of radioactive waste, dispersed throughout a multitude of international places and communities. These sites include the "atomic spaces" (Hales 1997) of the original Manhattan Project, the broader "atomic West" (Hevley and Findlay 1998)—especially the "tainted desert" lands in Nevada, Arizona, and California (Kuletz 1998)—and global "nuclear wastelands" (Makhijani, Hu and Yih 1995), all colonized by nuclear weapons production and testing activities. LeBaron (1998) provides a more sweeping version of the nuclear legacy trope, using it to encompass environmental and health effects, secret government medical experiments, civilian and military accidents, ongoing stockpiling of weapons by the original nuclear nations, prospects for proliferation to additional nations, and the growing specter of nuclear terrorism. Masco (2006) extends the trope to the cultural domain, describing the nuclear weapons enterprise as a "national fetish" (20) and the "central American project of the twentieth century" (25). Paradoxically, he also notes that Cold War culture was marked by a "near erasure of the nuclear economy from the public view" and a "banalization of the U.S. nuclear weapons in everyday American life" (4–5). This cultural masking of the ubiquitous "atomic bomb system" (Arney, 1991) continues, largely unchallenged, in the post–Cold War era (Taylor et al. 2005).

Gerber (2002) invokes "the Cold War legacy" of one nuclear site, the Hanford reservation, to address health effects controversies, management failures, and official secrecy associated with its operations. After that secrecy broke down dramatically in the 1980s, at Hanford and throughout the U.S. nuclear weapons production complex, the legacy trope appeared prominently in analyses by the Congressional Office of Technology Assessment (USOTA 1991) and in a series of publications associated with the Department of Energy's 1990s "openness initiative" (USDOE 1995, 1997). Those publications highlighted additional themes of complexity, environmental and economic costs, and the urgent need for concerted action by officials and stakeholders. Most recently, a report on the closure of the DOE's Rocky Flats site (USDOE 2006) enacts a new rhetorical move, reframing that project's legacy as one of *knowledge* acquired by a virtuous learning organization in its pursuit of a safe, cost-effective cleanup. The image of irresponsible weaponeering, supplanted in the 1990s by that of trustworthy environmental stewardship, now fades further from public memory and attention.

"Legacy," then, is an especially apt term for our purposes, because it ties together the past, present, and future dimensions of the nuclear condition. Legacies are historically produced and inherited from the past, confronted and remade in the present, and bequeathed to future generations. While the materiality of this process is undeniable, this volume uniquely explores its symbolic dimensions—the ways in which discourse operates to shape the expression and interpretation of nuclear legacies. Accordingly, temporality is a recurring theme of this discussion, reflected in the historicity of nuclear decisions, the urgency of the problems currently faced by nuclear officials and stakeholders, and the longevity of the products of nuclear operations. Space, or place, is another recurring theme, manifested in the settings for nuclear activities, the dissemination of nuclear materials throughout the environment, and the spatial mediation of responses to these conditions by the members of affected communities—for example, through their symbolization of spatial relations as risky proximity and safe remoteness. All of the chapters that follow grapple, in their own ways, with issues of time and space, and with the interrelated material and discursive phenomena that characterize particular nuclear scenes.

OVERVIEW OF THE VOLUME

The chapters collected in this volume variously address these "nuclear legacy" themes. Although they all take the analysis of communication as their central focus, the chapter authors include an anthropologist and two historians who integrate that focus with their own disciplinary concerns. As a result, the chapters offer the fruits of both a discipline-specific and an interdisciplinary focus on discursively-constituted relations between knowledge, power, culture, and institutions in the post–Cold War nuclear weapons production complex. As a result, the chapters speak to the interests of a number of communication subfields such as political, environmental and organizational communication and rhetoric, as well as to larger audiences seeking new models and resources for studying this site.

Part One: The Discourse of Officials and Stakeholders of Nuclear Weapons Production

The volume's first set of studies focuses on the role of communication in shaping relationships among and between the members of various groups associated with the nuclear weapons production complex. Specifically, these studies illustrate how the discourses developed by groups at particular facilities help their members to make sense of nuclear weapons production, and to orient their participation in and around that process (e.g., as workers, community activists, etc.). These discourses, the authors argue, construct particular meanings for the causes, figures, technologies, historical events, and consequences of nuclear weapons production. Additionally, they provide their speakers with criteria for evaluating multiple and conflicting knowledge claims surrounding those phenomena, and for producing what they view as legitimate and authoritative claims in associated controversies. As a result, we may understand the conflicts within and between these groups as arising from the multiple and competing discourses which they employ, and the unequal power relationships through which they have been able to position some discourses as superior to others. As these studies indicate, this conflict is potentially exacerbated in the relationships between the discourses of elite technical experts and community stakeholder groups engaged in assessing the risks associated with nuclear weapons production. This set of studies, then, examines the social, political, economic, and environmental consequences that emerge from particular, situated interaction between the officials and stakeholders of nuclear weapons production.

Our first study focuses on issues arising from activities at the Fernald, Ohio, production facility. In her essay, "Convergence and Divergence in the Public Dialogue on Nuclear Weapons Cleanup," Jennifer Duffield Hamilton reports on her study of a nineteen-year-long public dialogue at this site, and assesses more broadly the communication occurring in mandated forums for nuclear public participation that have expanded in the past decade. Specifically, she argues that a trend toward dialogue and consensus-building in these forums has enabled officials and stakeholders to practice mutual influence and achieve common perspectives on nuclear weapons cleanup. Her analysis examines how stakeholder groups initially constructed the problems and solutions of environmental cleanup in a climate of grievance and mistrust. An increasing reliance on dialogue, however, enabled the affected groups to collaborate in making decisions about site cleanup and waste disposal. This period was followed by emerging conflict surrounding the resolution of future land use and long-term stewardship at the site. Importantly, Fernald stakeholders were able to continue in dialogue throughout these stages, despite enduring conflicts.

Hamilton's study is significant on at least four levels. First, it provides a rare qualitative and longitudinal focus on concrete interaction between participants in these settings, as opposed to abstract characterizations of contextual factors. As a result, it enhances our basic understanding of this communication, a concern that often eludes interests bent on assessing and "fixing" it. Second, the study indicates the complexity of group and organizational identities operating in nuclear weapons cleanup: Here, participants were able to initially generate a working consensus partly because *site* DOE officials affiliated with local stakeholders against potentially obstructionist officials at DOE *headquarters*. Third, it provides a mixed reply to recent challenges made to the value of employing normative ideals in public participation programs (discussed above). On the one hand, Hamilton argues that normative ideals operating at the outset of group deliberations produced identification with the process that sustained participants through a period of substantive conflict. On the other hand, the groups did not practice all of the normative ideals derived from critical theory, suggesting that local flexibility and improvisation are equally important to producing successful outcomes. Additionally, the study confirms that both cooperation *and* conflict may contribute to successful dialogue, and that dialogue does not necessarily or permanently resolve to a single end-state. These participants answered the existential question of how they should live together with nuclear contamination by keeping the conversation going, despite their occasional moments of disappointment and discomfort. Finally, while this study presents something of a "success story," it also indicates how the possibilities for that success are currently contained by the DOE around issues of environmental cleanup. Although the DOE enforces a rigid boundary between those issues and others associated with national security and weapons production, other studies in this volume critique the arbitrary and porous nature of those boundaries.

Shifting to one of the original sites of nuclear weapons production, William Kinsella and Jay Mullen examine another protracted conflict in their chapter,

"Becoming Hanford Downwinders: Producing Community and Challenging Discursive Containment." They focus specifically on the role of communication in the emergence of the Hanford regional "downwinder" community, and in the downwinders' populist challenge to the secretive and technocratic frames that have historically dominated public discourse about operations at Hanford. Their study traces key moments in the history of this stakeholder group, beginning with the initial recognition by some neighbors of Hanford of a troubling pattern of health problems among regional residents and former residents. Linking their concerns with those of investigative journalists and community activists, these residents helped motivate an unprecedented series of disclosures by Hanford's managers regarding the environmental impacts of past operations. Those revelations pointed not only to a legacy of environmental contamination and health hazards, but also to a prolonged disregard of the implicit compact between citizens and government officials. These material and discursive legacies became themes for the production of a shared downwinder identity. Public reaction to these revelations, within and beyond the downwinder community, in turn inspired government responses including a pair of major health effects studies (the Hanford Environmental Dose Reconstruction Project and the Hanford Thyroid Disease Study), and the creation of a health information network and archives. Both critics and defenders of Hanford operations have subsequently looked to these programs for validation in policy disputes and legal proceedings. The authors examine the subsequent communication dilemmas confronting the program sponsors, administrators, researchers, and downwinder community members as they have negotiated and renegotiated boundaries that shape "reliable" and "truthful" knowledge about Hanford and its health effects.

At the most general level of analysis, this chapter provides a case study of the production of community through the development of a shared narrative and an associated, shared identity. The case study links a number of approaches drawn from organizational and group communication theory, examining processes of rhetorical identification, sensemaking, narrativity, and symbolic convergence. Drawing as well on Beck's model of contemporary "risk society" and on Luhmann's concept of "ecological communication," the authors argue that material hazards and failures of democracy are co-constituted through the downwinders' discursive efforts to make sense of their shared situation. This perspective provides a basis for illuminating, and potentially enhancing, interactions among citizen stakeholders, public officials, and technical specialists who engage with each other to negotiate the implications of Hanford. As part of that analysis, the authors further elaborate a model of "public expertise" (Kinsella 2004) relevant to broader issues of environmental conflict and democratic process. At the conclusion of their chapter, the authors return to issues of organizational communication by commenting on the implications of their case as a study in organizational crisis and change, and reflect on its implications for nuclear history, memory and heritage.

We end this section with an essay by Eric Morgan, who utilizes the ethnography of communication tradition to examine "Regional Communication and Sense of Place Surrounding the Waste Isolation Pilot Plant." Since 1999, the Waste Isolation Pilot Plant (WIPP), located near the city of Carlsbad in southeastern New Mexico, has stored "transuranic" waste from the Department of Energy's defense programs. As the nation's first underground facility designed to permanently store radioactive waste, the WIPP site has been the focus of numerous discourses among New Mexico groups for over twenty years. Within these discourses, Morgan argues, distinctly cultural premises can be heard that express how the residents of southeastern New Mexico can and should live in relation to the nuclear land. Utilizing cultural discourse theory, he tracks the range of meanings that the WIPP site has acquired for various inhabitants of the region. His argument focuses on the meaning of the WIPP site for those who live in Carlsbad and surrounding communities, and on how their local and regional sense of place changed with the opening of WIPP in 1999. Morgan argues that while the image of "isolation" dominates public discourse surrounding radioactive waste stored at WIPP, that image is confounded on two levels: at the level of validity, in that the ability of waste storage sites to successfully contain radioactive waste over long periods of time is a subject of enduring controversy; and at the level of metaphor, in that radioactive waste is clearly not—and should not be—isolated from discourse, which it compels as a medium of managing its associated hazards. In analyzing the discourse of local residents, Morgan finds two codes which they employ to rationalize the development of WIPP and assimilate its operations within their existing cultural frameworks: A code of *economy* which depicts WIPP as a necessary and valued source of reliable employment, and a spur to regional economic development, which has "rescued" the region following the decline of its mining industry; and a code of *nothingness* which depicts the place of WIPP's location as remote, desolate, and otherwise valueless. Morgan notes a potential conflict between these two themes arising from their simultaneous assertion of both value and valuelessness. In this way, Morgan suggests, the developers of nuclear sites must first evacuate them of existing function for local residents so as to successfully assert a value surrounding the location of radioactive waste disposal sites which justifies the local assumption of associated risk. The value derived from the site is reified, naturalized and normalized in a manner that precludes dissent.

Morgan's analysis offers two benefits to readers. The first is that it examines one region's communication practices regarding a situation that is emerging as a serious concern for other communities surrounding proposed nuclear waste storage sites (e.g., Las Vegas and Salt Lake City). Here, scholars can build from this study by comparing and contrasting regional nuclear discourses, and their role in alternately accommodating and opposing proposed sites (see also Glass 1993; Kuletz 1997). Second, in using ethnographic practices and cultural discourse theory, Morgan provides researchers with a model for systematically investigating the form and content of actual speech by stakeholders as they interpret these situations, and respond in ways that formulate and express their interests. By applying the ethnography of communication tradition and cultural discourse theory to nuclear communication phenomena, this chapter suggests the possibilities for future use of these resources.

Part Two: Organizing the Past, Present and Future of Nuclear Weapons Production

Our second set of studies focuses on the discourses employed by the officials and stakeholders of nuclear weapons production as they meaningfully configure the relationships between the past, present, and future of this enterprise. While this type of discourse occurs within all organizations and communities, it is one that is uniquely volatile and significant for the process of nuclear weapons production. This is true for at least two reasons. First, and perhaps most importantly, nuclear weapons possess a primordial power to shatter the organic continuity that is traditionally associated with unfolding periods of time. The phenomenology of terror associated with the Cold War nuclear arms race was partly grounded in popular fear not only of one's own individual death resulting from a nuclear exchange and the death of one's larger social group, but also the potential, permanent extinction of all humanity. As a result, discourse about nuclear weapons is uniquely shadowed by issues of mortality and continuity, and the temporal dimensions of narratives surrounding nuclear weapons production thus serve as registers of ideological argument favoring or opposing the continuity of this enterprise. The second reason is perhaps more tangible, but no less symbolic. It involves the process by which identifiable historical periods of operations in the nuclear weapons production complex are meaningfully characterized and configured—for example, as either continuous or discontinuous, and as repetition or rupture (or innovation). These narrative punctuations of historical activity can be controversial in that they identify "lessons" of the past, and how those lessons inform the present and future of nuclear weapons production. As such, these narratives facilitate the preferred arguments of various groups in the tense and unstable eras following the end of the Cold War and the events of 9/11.

Our first study in this section is Jason Krupar and Stephen Depoe's "Cold War Triumphant: The Rhetorical Uses of History, Memory, and Heritage Preservation within the Department of Energy's Nuclear Weapons Complex." They argue that, since the end of the Cold War, Americans have sought to grasp the significance and consequences of their country's decades-long struggle for global power with the Soviet Union. In this process, various official and unofficial narratives contend for legitimacy and authority over the form and content of that history. As sites that vividly objectify those narratives (e.g., as educational exhibits), museums form a principal locus in this cultural struggle. In recent years, a narrative master frame has emerged that tells the story of a triumphant United States who "won" the Cold War in part because of the industrial and military might associated with its nuclear weapons arsenal. This triumphalist frame has influenced not only U.S. foreign policy and military strategy, but also the way in which government officials and local stakeholders have attempted to preserve and present the multiple legacies of nuclear weapons production over a fifty-year period. This chapter examines the influence of the "Cold War triumphalism" frame in the development of heritage preservation and museum development efforts at three locations across the weapons complex. These sites include the opening of the Atomic Testing Museum in Las Vegas, Nevada; the discussion of a possible multi-use education facility and historical exhibit at the former weapons grade material production facility at Fernald, Ohio; and the struggle to launch a Cold War museum at the former production facility at Rocky Flats, located outside Denver, Colorado. The cases illustrate the power of the triumphalist narrative frame as a form of "discursive containment" (Kinsella 2001) in nuclear historical controversy (see also Taylor and Freer 2002). Heritage preservation efforts that emphasize the "positive" role played by nuclear weapons production in America's Cold War experience have received more political support and government funding than efforts that strive for a more balanced, multi-vocal or critical presentation of the past. The chapter ends with an examination of the Rocky Flats Virtual Museum, a web site that presents a harrowing tale of an industrial accident at the site in 1969 that threatened the city of Denver. This web site illustrates the rhetorical possibilities of local, unofficial historical narrative as a critique of official nuclear-heritage preservation efforts.

In the following chapter, "TRUTH is Generated HERE: Knowledge Loss and the Production of Nuclear Confidence in the Post–Cold War Era," Laura A. McNamara draws on six years of ethnographic research at the Los Alamos National Laboratory to document the impact of the 1992 U.S. testing moratorium on the social reproduction of knowledge among nuclear weapons designers. This event, she argues, produced interrelated technical and cultural consequences. At one level, the Cold War–era testing program enabled designers to produce an enormous body of knowledge about nuclear weapons. It also served, however, as an engine for reproducing and integrating the complex, tacit "ways of knowing" that are essential for designing and testing those weapons. Faced with the loss of the testing program, senior weapons experts have spent the past decade working to ensure that novice weaponeers can acquire the experientially-based practices and understandings that are required to make sound judgments about the state of the nation's nuclear stockpile. At the same time, the entire nuclear weapons design community is scrambling to develop, verify and validate new computationally based "ways of knowing" nuclear devices. Throughout, McNamara explores how weaponeers use discourse to understand and promote this work as a moral imperative worthy of being sustained into the future. In this way, their technical knowledge is discursively articulated with larger social and political bodies of knowledge that legitimate the development of nuclear weapons.

McNamara's study offers four benefits for observers of the nuclear weapons complex. First, it provides a rare ethnographic portrait of nuclear elites during a revealing period of traumatic change in which their conventional understandings and practices are in flux. As such, it depicts the alternating pride, creativity, frustration, despair, and persistence of nuclear professionals as they respond to the threatened extinction of their craft. Readers may find this response poignant, reassuring or worrisome, according to their own interpretive frameworks. Second, the study updates and responds to existing arguments concerning the potential "uninvention" of nuclear weapons in the post–Cold War era (Mackenzie and Spinardi 1995). Here, McNamara documents how nuclear officials have responded to the threat of lost designer knowledge by developing institutional programs that archive and sustain that knowledge. Although imperfect, these programs have functioned not only to document the history of nuclear weapons design but also to stimulate its future by recruiting and socializing new generations of designers. Third, McNamara models successful interdisciplinary study of this site by applying community of practice theory in ways that integrate the concerns of cultural anthropology and communication. This theory, she argues, "broadens the discussion of tacit-versus-explicit knowledge in science by linking the dynamics of knowing to the ongoing reinscription of shared identity, as expressed in a community's discursive products and through the practices of its members." McNamara subsequently combines this resource with additional research in organizational communication to critique the persistent "cryogenic" approach that dominates Knowledge Preservation programs among nuclear weapons designers. This approach manifests in a misleading metaphor of knowledge as *commodity* that continually mistakes the artifacts of knowledge for the organic social activities through which they are produced, circulated, and consumed. Finally, there is much for observers to consider here concerning the unfolding future of nuclear weapons development. McNamara documents several dilemmas facing designers, including determining whether their knowledge can and should be preserved, and when and how those activities threaten the fragile CTBT regime. For supporters of that regime, however, McNamara's study ends ominously: Nuclear weapons designers are currently developing a research program that *exploits* the uncertainty characterizing their operational environment, and whose framing of mission may ultimately succeed in "requiring" a resumption of explosive nuclear testing. McNamara's depiction of the role of discourse in these developments portends the discourse of CTBT promoters and opponents as that regime evolves.

The volume's final study is Bryan Taylor's "(Forever) At Work in the Fields of the Bomb: Images of Long-Term Stewardship in Post–Cold War Nuclear Discourse." This chapter examines official and stakeholder discourse surrounding the development of systems at nuclear weapons production and waste storage sites that are intended to protect human health and the environment for the duration of threats posed by contamination. At some sites, significantly, that duration involves tens—if not hundreds—of millennia. Taylor considers the ethics and politics of this discourse in several overlapping contexts, including: the general history and polysemy of stewardship discourse; the appropriation of that discourse to deliberate nuclear issues; the shadowing of ostensibly benevolent nuclear "stewardship" by an authoritarian tradition of "guardianship"; the development by officials and stakeholders of ideals for using this discourse in the post–Cold War era, and the inevitable challenges they faced in achieving these ideals. In this climate, Taylor argues, long-term stewardship (hereafter, LTS) discourse has formed a site of struggle between groups as they negotiate their interests in managing the environmental legacies of nuclear weapons production. Specifically, Taylor focuses on four themes in key official and stakeholder texts: the nature and consequences of discursively-constructed relationships between LTS and "cleanup" at affected sites; the consequences of "domestic" imagery in LTS discourse for the development of mutual trust between participants; the relative opportunities for public participation—and thus for achieving predictable, flexible, and accountable operations—created by LTS discourse; and finally, the consequences for the credibility and viability of LTS programs created by their discursive situation in a vast and disorienting temporal scale. Taylor concludes his study by considering obscured relationships between the discourses of "stewardship" operating in the DOE's environmental programs and in its weapons production programs (e.g., "stockpile stewards;" see the chapter by McNamara in this volume). Those two discourses, Taylor argues, are arbitrarily segregated, and "must be considered simultaneously if nuclear stewardship is to ethically serve the interests of dialogue and democracy."

Taylor's study serves the goals of this volume in three ways. First, it defamiliarizes a technocratic discourse circulating in the post–Cold War nuclear scene. By placing this discourse in its historical and cultural context, Taylor reveals that its dominant presentation of scientific, technical, bureaucratic, and legal issues is underwritten by a larger struggle for control over the development of nuclear weapons between nuclear citizens and officials. "Long-term stewardship" may be understood, as a result, as a current chapter in an ongoing saga. Secondly, Taylor models Alevsson and Karreman's (2000) "long-range" approach to investigating large-scale orders of discourse that mediate the understandings and interactions of organizational members and stakeholders. This account of LTS discourse documents both the structural conditions that constrain those activities, and also the contradictions, paradoxes and ambiguities that create moments and spaces for its autonomous development. Finally, Taylor resists persistent official efforts to segregate issues of nuclear weapons development and environmental cleanup (Kinsella 2005a, 62). By restoring their connections, Taylor offers a vision of how nuclear deliberations might ideally be conducted to serve both present and future generations.

Part Three: Critical Response

The volume concludes with a critical response that assesses its value for both scholarly inquiry and public participation in contemporary discursive struggles. In this response, Tarla Rai Peterson examines how scholars of environmental communication, organizational communication, and rhetoric might use and build from studies of nuclear controversy to advance ethically and politically engaged research programs. Specifically, she concludes, these studies offer strategies for transcending false and persistent oppositions that plague current theorizing and research (e.g., between the material and the symbolic, human and nonhuman, etc.). Additionally, they help us to envision a radically democratic culture characterized by inclusive dialogue and indeterminacy that resists strategic distortion and premature closure of discourse. Our cultural learning and growth associated with environmental politics, as a result, is enhanced.

CONCLUSION

We conclude our overview of this volume by noting a number of parallels between the present moment and the 1980s, the decade in which the nuclear criticism movement emerged. That decade was marked by a convergence of ominous trends including an aggressive posture adopted by the U.S. administration with respect to international affairs, proposals for new military technologies with the potential to exacerbate international tensions, and exhortations to U.S. citizens to defer domestic concerns and public debate in the context of a pressing national security threat. Despite the declared "end" of the Cold War, these trends appear to have re-emerged in the present moment, although they have done so in ways that exhibit both novelty and continuity. Novelty and contrast are evident in some aspects of the nuclear condition as it is constructed through discourses of terrorism, rogue statehood, and a mandate for preventative military action by U.S. forces. While each of these elements was present in Cold War–era nuclear discourse, they were configured in significantly different ways.[15] Alternately, continuity and similarity are manifest in the sustained, if not amplified, presumption of the nation-state regarding its legitimacy and authority in developing nuclear weapons (now ominously echoed in Iran's rhetoric), in the official promotion of military and technological solutions to political problems, and in the predominant appearance of apathy and inertia in public response to nuclear policy. In this dialectic of the new and the familiar, legacies of the Cold War nuclear moment structure public perceptions, policy agendas, and the discursive constructions of possible futures.

We must also note, however, that it was during the 1980s that some of the most effective responses to Cold War nuclearism emerged. Accordingly, the chapters that follow include narratives of (partially) successful citizen engagement with nuclear institutions, recognition (sometimes resigned and tentative), if not endorsement, by technocratic officials of the necessity of public participation in policy decisions, and a (limited and possibly temporary) reconfiguration of the nuclear weapons production complex to address environmental and public health, as well as national security, concerns. We believe that those narratives convey important lessons. The chapters also identify continuing and emerging sites of struggle, where these lessons might be put to use by motivated political actors. If that can be done, perhaps one legacy of the preceding nuclear era will be an expanded potential for realizing the vision developed by Schiappa (1989): a vision of the public as informed and active co-creators, along with officials and experts, of the evolving nuclear world.

NOTES

1. Access to heavy water facilitates plutonium production using natural, rather than enriched, uranium reactor fuel.

2. The complexity of the international system is illustrated by recent proposals for the sale of U.S. reactor technology to India, for storage in Russia of spent nuclear fuel from international sources, and for collaboration with Iran as an alternative to unilateral nuclear development.

3. Although this term is cumbersome, it is essential to clarifying the scope of our analysis. It signals that, for the most part, we address issues related to nuclear weapons, as distinguished from commercial nuclear power and other applications such as nuclear medicine. For the most part, we deal with issues of weapons production, rather than the military strategies surrounding their deployment and use. In focusing on the weapons production complex, we consider the network of physical sites and social scenes dispersed throughout the *United States* that support these operations. As demonstrated in the chapters that follow, that complex is dauntingly vast, dispersed and multifaceted, both as an organizational/institutional scene and as a setting for communication processes and practices.

4. This term is used broadly throughout this volume to include not only oral speakers, but also writers, performers, artists, photographers, film makers, and other symbol users.

5. Although it is not grounded in Bakhtinian dialogism, see the excellent related study by Krasniewicz (1992).

6. For example, Dauber (1993, 15, n. 8) grounds her study of nuclear argumentation in the dramatistic tradition of Symbolic Interactionism.

7. The dramatistic analysis of nuclear framing is strongly affiliated with the larger paradigm of social constructionism (see Benford 1993).

8. Fantasy theme analysis (Foss and Littlejohn 1986) is arguably influenced by dramatism. See also Bormann, Cragan and Shields (1996).

9. For the former, we nominate psychoanalytic (Taylor 1993a) and mythic criticism (Rushing 1986); for the latter, post-colonialism (Bjork 1995; see also Kuletz 1998).

10. For examples, see Williams (1988) for an integration of dramatism and deconstruction; Ausmus (1998) for a study that combines dramatism, critical theory, and argumentation theory to examine "nuclear winter" as both a metaphor and a condensed policy argument that is contested across discourse spheres; Mehan and Wills (1988) for a combination of dialogism and dramatism to study the discourse of antinuclear social movements; Meyer (1995) for a characterization of dominant Cold War frames for representing nuclear policy and antinuclear protest, and an examination of how those frames contended within three discourse "venues" (analogous to spheres) of presidential administrations, strategic experts, and scientific experts; and finally, Grey (2002) for a combination of dramatism and post-structuralism in deconstructing the evolving politics of representation in literature surrounding survivors of the Japanese atomic bombings (*hibakusha*).

11. While this claim emphasizes differences in disciplinary framing of the nuclear condition, the "see also" citations listed throughout this section refer to work completed by nuclear scholars in these disciplines that directly and indirectly engages nuclear discourse.

12. As opposed to re-allocating their warheads to "reserve" categories, thus creating the appearance of reduction but leaving those warheads available for use.

13. Although space does not permit full elaboration of these connections, the revival of nuclear weapons production by the United States affects the conditions of nuclear terrorism and proliferation in at least four ways: 1) by directly increasing the numbers of nuclear warheads and amounts of fissile materials that may be vulnerable to unauthorized diversion and use; 2) by indirectly increasing these numbers and amounts through stimulating the resumption of competitive nuclear weapons production by other nuclear states; 3) by enlisting the productivity of U.S. civilian nuclear power reactors in ways that erode existing arms control agreements (Bergeron 2002); and 4) by increasing existing resentment among other signatories to the NPT concerning apparent U.S. hypocrisy towards its obligatory pursuit of nuclear disarmament. This resentment may lead the agents of those nations to actively or passively facilitate the illegal diversion of nuclear weapons, weapons production technologies, and/or fissile materials to prospective nuclear nations and terrorists (Allison 2004; Langewisesche 2005).

14. Additional significant activities took place at academic sites including Columbia University, the University of Chicago's Metallurgical Laboratory, and the University of California's Berkeley Radiation Laboratory.

15. For example, as a participant in Cold War–era NATO, the United States did not forswear first use of tactical nuclear weapons in response to an overwhelming invasion of Western Europe by Warsaw Pact conventional forces. That form of first use, however, differs dramatically from other forms currently endorsed by U.S. policy—including a *preventive* attack by the United States (potentially employing nuclear weapons) on a sovereign nation believed to be *developing* a nuclear *capability* to attack the United States.

BIBLIOGRAPHY

Allison, Graham. 2004. *Nuclear terrorism: The ultimate preventable catastrophe*. New York: Times Books.

Alvesson, Mats and Dan Karreman. 2000. Varieties of discourse: On the study of organizations through discourse analysis. *Human Relations* 53: 1125–1149.

Arney, W. R. 1991. *Experts in the age of systems*. Albuquerque: University of New Mexico Press.

Ausmus, William A. 1998. Pragmatic uses of metaphor: Models and metaphor in the nuclear winter scenario. *Communication Monographs* 65: 67–82.

Bazerman, Charles. 2001. Nuclear information: One rhetorical moment in the construction of the information age. *Written Communication* 18: 259–295.

Beckman, P. R., S. P. Lee, M. N. Dobkowski, L. Campbell, and P. W. Crumlish. 1992. *The nuclear predicament: Nuclear weapons in the Cold War and beyond*. Englewood Cliffs, N.J.: Prentice Hall.

Benford, Robert D. 1993. Frame disputes within the nuclear disarmament movement. *Social Forces* 71: 677–701.

Benson, Thomas W., and Carolyn Anderson. 1990. The ultimate technology: Frederick Wiseman's *Missile*. Pp. 257–283 in *Communication and the culture of technology*, ed. Martin J. Medhurst, Alberto Gonzalez, and Tarla Rai Peterson. Pullman: Washington State University Press.

Bergeron, Kenneth D. 2002. *Tritium on ice: The dangerous new alliance of nuclear weapons and nuclear power*. Cambridge, Mass.: MIT Press.

Bjork, Rebecca S. 1992. *The strategic defense initiative: Symbolic containment of the nuclear threat*. Albany: SUNY Press.

———. 1995. Public policy argumentation and colonialist ideology in the post–Cold War era. Pp. 211–236 in *Warranting assent: Case studies in argument evaluation*, ed. Edward Schiappa. Albany: SUNY Press.

Bormann, Ernest G., John F. Cragan, and Donald C. Shields. 1996. An expansion of the rhetorical vision component of the symbolic convergence theory: The Cold War paradigm case. *Communication Monographs* 63: 1–28.

Broad, William J. 2003. Chain reaction. *The New York Times*, 3 August, sec. 4, pp. 1, 12.

Brummett, Barry. 1989. Perfection and the bomb: Nuclear weapons, teleology, and motives. *Journal of Communication* 39: 85–95.

Caldicott, Helen. 2002. *The new nuclear danger: George W. Bush's military-industrial complex*. New York: New Press.

Chaloupka, William. 1992. *Knowing nukes: The politics and culture of the atom.* Minneapolis: University of Minnesota Press.

Cheney, G. 1991. *Rhetoric in an organizational society: The management of multiple identities.* Columbia: University of South Carolina Press.

Cohn, Carol. 1987. Sex and death in the rational world of defense intellectuals. *Signs* 12: 687–718.

Dauber, Cori. 1988. Through a glass darkly: Validity standards, and the debate over nuclear strategic doctrine. *Journal of the American Forensic Association* 24: 168–180.

———. 1993. *Cold War analytical structures and the post- post-war world: A critique of deterrence theory*. Westport, Conn.: Praeger.

Dayton, David. 2002. Evaluating environmental impact statements as communicative action. *Journal of Business and Technical Communication* 16: 355–405.

Deetz, Stanley. 1991. *Democracy in an age of corporate colonization: Developments in communication and the politics of everyday life*. Albany: SUNY Press.

———. 1995. *Transforming communication, transforming business: Building responsive and responsible workplaces.* Cresskill, N.J.: Hampton Press.

Derrida, Jacques. 1984. No apocalypse, not now (full speed ahead, seven missiles, seven missives). *Diacritics* 14: 20–31.

Dowling, David. 1987. *Fictions of nuclear disaster*. Iowa City: University of Iowa Press.

Farrell, Thomas B., and G. Thomas Goodnight. 1981. Accidental rhetoric: The root metaphors of Three Mile Island. *Communication Monographs* 48: 271–300.

Fisher, Walter R. 1984. Narration as a human communication paradigm: The case of public moral argument. *Communication Monographs* 51: 1–22.

Foss, Karen A., and Stephen W. Littlejohn. 1986. *The Day After*: Rhetorical vision in an ironic frame. *Critical Studies in Mass Communication* 5: 317–336.

Futrell, Robert. 2003.Technical adversarialism and participatory collaboration in the U.S. chemical weapons disposal program. *Science, Technology, & Human Values* 28: 451–482.

Gabel, Josiane. 2004/2005. The role of U.S. nuclear weapons after September 11. *The Washington Quarterly* 28: 181–195.

Gerber, Michele S. 2002. *On the home front: The Cold War legacy of the Hanford nuclear site*. Lincoln: University of Nebraska Press.

Glass, Matthew. 1993. *Citizens against the MX: Public languages in the nuclear age*. Urbana: University of Illinois Press.

Goodnight, G. Thomas. 1982. The personal, technical, and public spheres of argument: A speculative inquiry into the art of public deliberation. *Journal of the American Forensic Association* 18: 214–227.

———. 1988. Argumentation and the nuclear age. *Journal of the American Forensic Association* 24: 141–142.

Grey, Stephanie Houston. 2002. Writing redemption: Trauma and the authentication of the moral order in *Hibakusha* literature. *Text and Performance Quarterly* 22: 1–23.

Gusterson, Hugh. 1996. *Nuclear rites: A nuclear weapons laboratory at the end of the Cold War*. Berkeley: University of California Press.

Hales, Peter B. 1997. *Atomic spaces: Living on the Manhattan Project*. Urbana: University of Illinois Press.

Hamilton, Jennifer Duffield. 2003. Exploring technical and cultural appeals in strategic risk communication: The Fernald radium case. *Risk Analysis* 2: 291–302.

Hardert, R. A., M. Reader, M. Scott, G. L. Moulton, and A. Goodman, 1989. A critical theory analysis of nuclear power: The implications of Palo Verde nuclear generating station. *Humanity and Society* 13: 165–186.

Hevly, Bruce, and John M. Findlay, eds. 1998. *The atomic West*. Seattle: University of Washington Press.

Hubbard, Brian. 1997. Rhetorical criticism after the Cold War. Master's thesis, Arizona State University.

———. 1998. Reassessing Truman, the bomb, and revisionism: The burlesque frame and entelechy in the decision to use atomic weapons against Japan. *Western Journal of Communication* 62: 348–385.

Hynes, Thomas J., Jr. 1988. You can't prove it here: Nuclear arms negotiation and the testing of evidence. *Journal of the American Forensic Association* 24: 155–167.

Kalaidjian, Walter. 1999. Nuclear criticism. *Contemporary Literature* 40: 311–318.

Kalinowski, Martin B. 2004. Nuclear arms races and arms control at the beginning of the 21st century. *Security Dialogue* 35: 217–225.

Katz, Steven B., and Carolyn R. Miller. 1996. The low-level radioactive waste-siting controversy in North Carolina. Pp. 111–139 in Carl G. Herndl and Stuart C. Brown, eds,. *Green culture: Environmental rhetoric in contemporary America*. Madison: University of Wisconsin Press.

Kauffman, Charles. 1989. Names and weapons. *Communication Monographs* 56: 273–285.

Kauzlarich, David, and Ronald C. Kramer. 1998. *Crimes of the American nuclear state: At home and abroad*. Boston: Northeastern University Press.

Keller, Bill. 2002. Nuclear nightmares. *The New York Times Magazine*, 26 May, pp. 22, 24–29, 51, 54–55, 57.

King, Andrew, and Kenneth Petress. 1990. Universal public argument and the failure of the nuclear freeze. *Southern Communication Journal* 55: 162–174.

Kinney, Aimee Guglielmo and Thomas M. Leschine. 2002. A procedural evaluation of an analytic-deliberative process: The Columbia River comprehensive impact assessment. *Risk Analysis* 22: 83–100.

Kinsella, William J. 1999. Discourse, power, and knowledge in the management of "big science": The production of consensus in a nuclear fusion research laboratory. *Management Communication Quarterly* 13: 171–208.

———. 2001. Nuclear boundaries: Material and discursive containment at the Hanford nuclear reservation. *Science as Culture* 10: 163–194.

———. 2004. Public expertise: A foundation for citizen participation in energy and environmental decisions. Pp. 83–95 in *Communication and public participation in environmental decision making*, ed. Stephen. P. Depoe, John W. Delicath, and Marie-France Aepli Elsenbeer. Albany: SUNY Press.

———. 2005a. One hundred years of nuclear discourse: Four master themes and their implications for environmental communication. Pp. 49–72 in *Environmental Communication Yearbook* 2, ed. Susan L. Senecah. Mahwah, N.J.: Lawrence Erlbaum Associates.

———. 2005b. Rhetoric, action, and agency in institutionalized science and technology. *Technical Communication Quarterly* 14: 303–310.

Krasniewicz, Louise. 1992. *Nuclear summer: The clash of communities at the Seneca Women's Peace Encampment*. Ithaca, N.Y.: Cornell University Press.

Kuletz, Valerie L. 1998. *The tainted desert: Environmental and social ruin in the American West*. London: Routledge.

Langewisesche, William. 2005. The wrath of Khan. *The Atlantic Monthly*, November, 62–68, 70–72, 74, 76–80, 82, 84–85.

Lawless, W. F., M. Bergman, and N. Feltovich. 2005. Consensus seeking versus truth seeking. *Practice Periodical of Hazardous, Toxic, and Radioactive Waste Management*, January: 59–70.

LeBaron, Wayne D. 1998. *America's nuclear legacy*. Commack, N.Y.: Nova Science Publishers.

MacKenzie, Donald, and Graham Spinardi. 1995. Tacit knowledge, weapons design, and the uninvention of nuclear weapons. *American Journal of Sociology* 101: 44–99.

Makhijani, Arjun, Howard Hu, and Katherine Yih. 1995. *Nuclear wastelands: A global guide to nuclear weapons production and its health and environmental effects*. Cambridge, Mass.: MIT Press.

Makhijani, Arjun. 2006, 26 June. Comments at Institute for Energy and Environmental Research curriculum development workshop, Durham, N.C.

Masco, Joseph. 2006. *The nuclear borderlands*. Princeton, N.J.: Princeton University Press.

McQuerry, Maureen. 2000. *Nuclear legacy: Students of two atomic cities*. Columbus, Ohio: Battelle Press.

Mechling, Elizabeth W., and Jay Mechling. 1991. The campaign for civil defense and the struggle to naturalize the Bomb. *Western Journal of Speech Communication* 55: 105–133.

Medvedev, Zhores A. 1990. *The legacy of Chernobyl*. New York: Norton.

Mehan, Hugh, Charles E. Nathanson, and James M. Skelly, J. 1990. Nuclear discourse in the 1980s: The unravelling conventions of the cold war. *Discourse and Society* 1: 133–165.

Mehan, Hugh, and John Wills. 1988. MEND: A nurturing voice in the nuclear arms debate. *Social Problems* 35: 363–383.

Mehta, Michael D. 1998. Risk and decision-making: A theoretical approach to public participation in techno-scientific situations. *Technology in Society* 20: 87–98.

Metzler, Maribeth S. 1998. Organizations, democracy, and the public sphere: The implications of democratic revolution at a nuclear weapons facility. *Communication Studies* 48: 333–358.

Meyer, David S. 1995. Framing national security: Elite public discourse on nuclear weapons during the Cold War. *Political Communication* 12: 173–192.

Mitchell, Gordon. 2000. *Strategic deception: Rhetoric, science, and politics in missile defense advocacy*. East Lansing: Michigan State University Press.

Nolan, Janne. 1999. *An elusive consensus: Nuclear weapons and American security after the Cold War*. Washington, D.C.: Brookings Institution Press.

Norris, Christopher. 1994. "Nuclear criticism" ten years on. *Prose Studies* 17: 130–139.

O'Neill, Dan. 1998. Alaska and the firecracker boys: The story of Project Chariot. Pp. 179–199 in *The atomic West*, ed. Bruce Hevly and John M. Findlay. Seattle: University of Washington Press.

Prosise, Theodore O. 1998. The collective memory of the atomic bombings misrecognized as objective history: The case of the public opposition to the National Air and Space Museum's atom bomb exhibit. *Western Journal of Communication* 62: 316–347.

Ratliff, Jeanne Nelson. 1998. The politics of nuclear waste: An analysis of a public hearing on the proposed Yucca Mountain nuclear waste repository. *Electronic Journal of Communication* 8, *www.cios.org/getfile\Ratliff_V8N198* (accessed March 23, 1999).

Rhodes, Richard. 1994. Atomic logic: The bomb in the post–cold war world. *Rolling Stone*, 24 February, 30–37, 69.

Roff, Sue R. 1995. *Hotspots: The legacy of Hiroshima and Nagasaki*. New York: Cassell.

Rushing, Janice Hocker. 1986. Ronald Reagan's "star wars" address: Mythic containment of technical reasoning. *Quarterly Journal of Speech* 72: 415–433.

Ruthven, Ken K. 1993. *Nuclear criticism*. Carlton, Vic.: Melbourne University Press.

Saleska, Scott. 1989. *Nuclear legacy: An overview of the places, problems, and politics of radioactive waste in the U. S.* Washington, D.C.: Public Citizen, Critical Mass Energy Project.

Santos, Susan L., and Caron Chess. 2003. Evaluating citizen advisory boards: The importance of theory and participant-based criteria and practical implications. *Risk Analysis* 23: 269–279.

Schelling, Thomas. 2006. The legacy of Hiroshima. Pp. 313–325 in Thomas Schelling, *Strategies of commitment and other essays*. Cambridge, Mass.: Harvard University Press.

Schiappa, Edward. 1989. The rhetoric of nukespeak. *Communication Monographs* 56: 253–272.

Schwartz, Stephen. 2001. The new-nuke chorus tunes up. *Bulletin of the Atomic Scientists*, July/August, 30–35.

Schwenger, Peter. 1992. *Letter bomb: Nuclear holocaust and the exploding word*. Baltimore: Johns Hopkins University Press.

Sherwin, Martin. 2003. *A world destroyed: Hiroshima and its legacies*. Stanford: Stanford University Press.

Skillington, Tracey. 1997. Politics and the struggle to define: A discourse analysis of the framing strategies of competing actors in a "new" participatory forum. *The British Journal of Sociology* 48: 493–513.

Taylor, Bryan C. 1990. Reminiscences of Los Alamos: Narrative, critical theory and the organizational subject. *Western Journal of Speech Communication* 54: 395–419.

———. 1992. The politics of the nuclear text: Reading Robert Oppenheimer's *Letters and Reflections*. *Quarterly Journal of Speech* 78: 429–449.

———. 1993a. Register of the repressed: Women's voice and body in the nuclear weapons organization. *Quarterly Journal of Speech* 79: 267–285.

———. 1993b. *Fat Man and Little Boy*: Cinematic representation of interests in the nuclear weapons organization. *Critical Studies in Mass Communication 10*: 367–394.

———. 1996. Make bomb, save world: Reflections on dialogic nuclear ethnography. *Journal of Contemporary Ethnography* 25: 120–143.

———. 1997a. Home zero: Images of home and field in nuclear-cultural studies. *Western Journal of Communication* 61: 209–234.

———. 1997b. Revisiting nuclear history: Narrative conflict at the Bradbury Science Museum. *Studies in Cultures: Organizations and Societies* 3: 119–145.

———. 1997c. Shooting downwind: Depicting the radiated body in epidemiology and documentary photography. Pp. 289–328 in *Transgressing discourses: Communication and the voice of other*, ed. Michael Huspek and Gary Radford. Albany: SUNY Press.

———. 1998a. Nuclear weapons and communication studies: A review essay. *Western Journal of Communication* 62: 300–315.

———. 1998b. The bodies of August: Photographic realism and controversy at the National Air and Space Museum. *Rhetoric and Public Affairs* 1: 331–361.

———. 2001. Nuclear communication studies. [Review of John Canaday, *The nuclear muse: Literature, physics, and the first atomic bombs*]. *The Review of Communication* 1: 81–86.

———. 2002. Organizing the "unknown subject": Los Alamos, espionage, and the politics of biography. *Quarterly Journal of Speech* 88: 33–49.

———. 2003a. "Our bruised arms hung up as monuments": Nuclear iconography in post-Cold War culture. *Critical Studies in Media Communication* 20: 1–34.

——. 2003b. Nuclear waste and communication studies [Review of Thomas V. Peterson, *Linked arms: A rural community resists nuclear waste*]. *The Review of Communication* 3: 285–291.

Taylor, Bryan C., and Brian Freer. Containing the nuclear past: The politics of history and heritage at the Hanford Plutonium Works. *Journal of Organizational Change Management* 15: 563–588.

Taylor, Bryan C., William J. Kinsella, Stephen P. Depoe, and Maribeth S. Metzler. 2005. Nuclear legacies: Communication, controversy, and the U.S. nuclear weapons production complex. Pp. 363–409 in *Communication Yearbook 29*, ed. Pamela Kalbfleisch. Mahwah, NJ: Lawrence Erlbaum Associates.

Teller, Edward. 1962. *The legacy of Hiroshima*. New York: Doubleday.

Tietge, David J. 2002. *Flash effect: Science and the rhetorical origins of Cold War America*. Athens: Ohio University Press.

Toker, Caitlin Wills. 2002. Debating "what ought to be": The comic frame and public moral argument. *Western Journal of Communication* 66: 53–83.

U.S. Department of Energy. National Nuclear Security Agency (USDOE). 2006, 28 June. NNSA establishes new office to lead future of nuclear weapons complex, *www.nnsa.doe.gov/docs/newsreleases/2006/PR_2006-06-28_NA-06-20.htm* (Accessed July 1, 2006).

U.S. Department of Energy. Office of Environmental Management (USDOE). 1995. *Closing the circle on the splitting of the atom: The environmental legacy of nuclear weapons production in the United States and what the Department of Energy is doing about it.* Washington, D.C.

U.S. Department of Energy. Office of Environmental Management (USDOE). 1997. *Linking legacies: Connecting the Cold War nuclear weapons production processes to their environmental consequences.* Washington, D.C.

U.S. Department of Energy. Office of Environmental Management (USDOE). 2006. *The Rocky Flats closure legacy: A lessons learned report*, rockyflats.apps.em.doe.gov/ (accessed September 29, 2006).

U.S. Congress. Office of Technology Assessment (USOTA). 1991. *Complex cleanup: The environmental legacy of nuclear weapons production.* Washington, D.C.

Von Meier, Alexandra., Jennifer Lynn Miller, and Ann C. Keller. 1998. The disposition of excess weapons plutonium: A comparison of three narrative contexts. *The Nonproliferation Review*, Winter: 20–31.

Wertsch, James V. 1987. Nuclear discourse. *Communication Research* 14: 131–138.

Williams, David Cratis. 1988. Nuclear criticism: In pursuit of a "politically enabling" deconstructive voice. *Journal of the American Forensic Association* 24: 193–205.

Winner, Langdon. 1980. Do artifacts have politics? *Daedalus* 109: 129–136.

Part One

THE DISCOURSE OF OFFICIALS AND STAKEHOLDERS OF NUCLEAR WEAPONS PRODUCTION

2

Convergence and Divergence in the Public Dialogue on Nuclear Weapons Cleanup

Jennifer Duffield Hamilton[1]

Public participation has been a central feature of environmental decision making since 1970; however, it was not until the late 1980s and throughout the 1990s that public participation became part of a rapidly changing landscape of nuclear weapons debate. The U.S. Department of Energy (hereafter, DOE), a once self-regulated agency with a national security mission, came under increasing scrutiny by the media and residents living near its nuclear weapons production facilities across the United States. During this time, citizens and regulators filed lawsuits and Congress investigated the DOE's noncompliance with environmental laws. Many DOE facilities were subjected to increasing environmental regulation that initiated decades-long cleanup projects and legal obligations to involve the public in those decisions (Fehner and Gosling 1996). Through public forums mandated by the Comprehensive Environmental Response, Compensation, and Liability Act (CERCLA) guiding the cleanup of many of these sites (Veiluva 1995), nuclear stakeholders began deliberating and contesting issues of environmental contamination and potential human health risks. These mandated public participation techniques, such as public review periods and hearings, have been criticized for eliciting citizen input after solutions have been proposed, inviting citizen comment without promoting adequate learning or discussion of the issues, and excluding underlying citizen values from the discussion (Applegate 1998; Fiorino 1989).

Due in part to the history of secrecy regarding contamination from nuclear weapons production and in part to rigid and confining participation forums prescribed by law, stakeholders such as the DOE, federal and state Environmental Protection Agencies (EPAs), and nearby residents and community groups have been entangled in conflict in these legally-mandated forums. In the early 1990s, citizen demands for openness and inclusion in the

DOE's environmental cleanup decisions led the agency to reach beyond traditional public hearing and comment-and-review periods and try forums that relied on dialogue and consensus building among stakeholders (Keystone Center 1996). These forums are part of a larger trend in public participation to seek solutions to environmental disputes through collaboration among various groups using formats such as workshops and citizen advisory boards (Crowfoot and Wondolleck 1990). The growing reliance on dialogic formats for participation together with DOE-community relationships that are increasingly long-term in nature, presents an opportunity to examine the evolution of stakeholder perspectives over time and in conjunction with new settings for public participation that emerged as the Cold War was ending.

In this chapter, I examine the interaction of stakeholder groups (DOE, U.S. and Ohio EPAs, and a community group called FRESH) at the DOE site in Fernald, Ohio, as they constructed the problems and solutions of nuclear weapons cleanup over a nineteen-year period. To analyze stakeholder perspectives, I rely on James Wertsch's (1987) categories of *particularistic* and *universalistic* which refer to the ways in which participants envision their connection to others in nuclear issues (i.e., as "us-versus-them" or "we're all in this together", respectively). I also make use of Wertsch's related categories of *de-contextualized* and *contextualized* frames, which refer to speakers' grounding of their nuclear arguments in either abstract or situated reasoning. *Convergence* was found, I will argue, when stakeholder groups operated out of the same perspective or frame for understanding events, and *divergence* when they did not. I explore how participants in deliberation at this site have used discourse grounded in these perspectives to inform and influence each other, and to create shifts in each other's understandings of nuclear weapons cleanup.

Specifically, I argue that a productive tension developed among these groups, including an evolving pattern of convergence and divergence of perspectives. Changes in the ways in which the FRESH, EPAs, and DOE groups understood the issues at hand were influenced by their ability to envision the local actors working together to achieve environmental cleanup. This vision of a "we" acting in concert at the local level was instrumental in transitioning their relationship to a certain degree of cooperation and convergence during initial decisions related to the cleanup budget and waste disposition. This was a significant achievement given the history of distrust and public skepticism regarding the plant's contamination of the environment. As decision-making subsequently progressed to consider long-term stewardship of the site, these shared understandings shifted and reformed in ways that envisioned site officials as affiliated with the interests of DOE Headquarters rather than with local interests in identifying future land use. Importantly, this case revealed that stakeholders were able to continue in dialogue despite their conflicts and in great part because of the common view of "public participation" they had achieved. When nuclear discourse among stakeholders is considered over time, this study suggests, dialogue emerges as a process that moves back and forth between convergence and divergence, rather than necessarily toward consensus or resolution of conflicting views as a natural endpoint. This finding thus challenges normalized views of dialogue in the public participation literature.

PUBLIC PARTICIPATION AND NUCLEAR WEAPONS STUDIES

This essay builds upon two areas of research increasingly connected by common concerns. The first is public participation literature and associated efforts to theorize criteria for democratic involvement in environmental decision-making. These studies are increasingly concerned with dialogue, learning, and consensus building as central goals for public participation. Traditional public participation techniques such as comment periods and public hearings have been highly criticized for an inability to involve citizens in ways that impact decisions, thus compounding contention among public participants and decision-makers (Killingsworth and Palmer 1992). Scholars and practitioners of public participation have recognized these limitations and have advanced a variety of models that attempt to improve traditional public participation. Several communication scholars have done so by drawing a contrast between a "one-way" transfer of information from experts to citizens and a "two-way" exchange of information and values among these groups (Juanillo and Scherer 1995; Katz and Miller 1996; Waddell 1996). This research calls for mechanisms that create early and regular interaction and allow participants to analyze and debate the issues in an open exchange (Daniels and Walker 1996; Fiorino 1990).

Several projects have offered criteria for fulfilling that goal. Fiorino (1989, 1990) has advanced a set of criteria based on democratic theories including direct participation, participation on an equal basis, face-to-face interaction over time, and the ability to "co-determine" policy with agency officials, that has been highly influential in this literature. Another important project in this respect is Renn, Webler, and Wiedemann's (1995) edited volume evaluating the degree to which forums such as regulatory negotiations and citizen advisory boards allow participants to deliberate issues through fair and competent processes, conceptualized as achieving Habermas' ideal speech situation and communicative competence (Habermas 1970, 1984, 1987). Some argue, alternately, that factors beyond criteria for democratic processes account for the outcomes of public participation (Chess and Purcell 1999). For example, participation based on group learning and analysis of the issues is emphasized in what Laird (1993) calls "participatory analysis" and Daniels and Walker (1996) term "collaborative learning." These approaches recommend a process in which participants are collaboratively "defining the problem and generating alternatives [which] makes for meaningful social learning as constituencies sort out their own and others' values, orientations, and priorities" (Daniels and Walker 1996, 73). Dialogue and group learning are envisioned as essential to moving beyond current limitations of public participation through providing opportunities to jointly identify, analyze, and debate problems and solutions.

Despite these developments, an important argument has emerged that participation should include an exploration of *differences* as much as similarities among the various interests. "[P]olicy processes often have a one-sided emphasis on 'common ground' and 'mutual understanding'. With such a view, however, diversity, difference, conflict, and incompatibility are at risk of getting lost in our understandings of dialogue and discourse" (Tuler 2000, 13). This argument emerged in part as a critique of consensus as the desired outcome of public deliberation. A majority of recent participation models are designed around Habermasian assumptions of the public sphere in which a community consensus emerges from deliberation among equal and rational citizens (Wills-Toker 2004). The notion of consensus has been questioned in part because of the implausibility of an exchange in which people are truly equal and power differences have been removed, and in part because of the tendency to suppress important differences in an effort to find common ground (Mouffe 1999; Santos and Chess 2003). Ingham (1996, 202) suggests a process she calls the "recognition of dissensus," arguing that when dialogue is thought of in this way, it is a process through which participants are "identifying with each other, by establishing stances in relationship to each other". The collaborative learning approach similarly envisions public participation as discovering and understanding the relationships among viewpoints (Daniels and Walker 1996). This essay extends this conversation through analysis of a sustained case of public participation in which similarity *and* difference, cooperation *and* conflict emerged as important to the process of dialogue. In doing so, I challenge normalized views of dialogue as naturally progressing toward agreement and assert the importance of conflict to the process of dialogue.

The second area of research this essay builds upon is nuclear communication studies and efforts to conceptualize controversies over nuclear weapons production, use, and environmental cleanup as a symbolic struggle over meaning (Taylor 1997). These studies explore sites of meaning production or places in which this cultural dialogue is conducted such as nuclear weapons organizations (Freer 1994; Gusterson 1996; Taylor 1990), opposition organizations (Ratliff 1997a), historical accounts (Prosise 1998; Taylor 1997; Taylor and Freer 2002), and films about the nuclear age (Taylor 1993). These studies have revealed practices and discourses that naturalize official understandings of nuclear weapons as well as strategies for contesting and offering alternative accounts of nuclear events. Importantly, these studies shed light on a continuing struggle among competing interpretations of nuclear issues in many facets of American culture. Two developments in this respect are relevant to this study.

First, nuclear scholars have found that the pro- and anti-groups involved in the nuclear arms debate were unable to overcome distinct ways of thinking and talking to develop common understandings of the nuclear situation (Gusterson 1996). Wertsch (1987, 102) explains:

> One of the most striking features of the nuclear arms debate is that representatives of various viewpoints seem to be incapable of communicating with one another . . . What we see are several different perspectives that are for all practical purposes monologues . . . Each of these perspectives presents a coherent world view . . . but each of them also is in fundamental contradiction with the others in what it sees as the major problem and how it can be solved.

Gusterson (1996) came to a similar conclusion in his study of the Lawrence Livermore National Laboratory, arguing that nuclear weapons scientists and antinuclear activists operated out of ideologies that were incapable of transforming one another. Gusterson (1996, 4) explains: "The nuclearist and antinuclear worldviews are both plausible constructions of the world that are unable to defeat one another." These studies have revealed an inability of interests competing to establish nuclear meanings to develop common interpretations. To this discussion, this essay contributes a case study in which nuclear stakeholders, despite maintaining differences, were able to achieve some common ways of interpreting and speaking about nuclear cleanup through the trend toward greater interaction and dialogue noted above. Significantly, these accomplishments proved sufficiently flexible and durable to accommodate subsequent disagreements.

Second, communication scholars have critiqued the culture operating throughout the DOE's nuclear weapons complex as marked by "containment" that discourages undistorted communication. Most notably, Kinsella (2001) has argued that the discourse of secrecy that prevailed during a Cold War era of national security has evolved into new forms of discursive containment revolving around claims to expertise. Ratliff (1997b) similarly observed the phenomenon of expertise containing debate at public hearings through speakers' assertion of the authority of official-technical discourses. Hamilton

(2003a) observed citizens and officials combining elements of technical and cultural rationalities in an effort to form connections and persuade one another. Still, Kinsella (2001) argues that in a post–Cold War era, the DOE began providing access to information about nuclear weapons production, but continues to control the meaning of that information and the range of speakers authorized to make that meaning. These advancements share the concern of public participation studies with interaction among government officials, experts, and citizens in the public sphere as well as the ability of participants to openly form meaning through uncontained communication.

The common concerns of these two lines of literature come together in a group of studies examining the DOE's public participation programs. These studies have examined public participation primarily through the efforts of DOE's site-specific advisory boards (SSAB) or local activist and community groups. Scholarly research has done so by assessing the democratic structures and processes inherent in specific participation formats, particularly the SSABs (Applegate 1998; Duffield and Depoe 1997; Williams 2002). These studies have documented various structural and process factors such as broad representation, openness, process fairness, influence on decisions, face-to-face communication over time, sustained focus, leadership and membership stability, technical support, effective cooperation, and member commitment as key to SSAB success. In addition to studies of the advisory boards, scholarly research has documented the emergence and relative effectiveness of activist groups at DOE sites in Hanford, Washington and Fernald, Ohio (Kaplan 2000; Ratliff 1997a; Sheak and Cianciolo 1993). Alternately, other studies depict a lack of activism and reveal forms of social control that promote the government's frame and suppress collective action (Cable, Shriver, and Hastings 1999).

"Evaluation research" has studied this scene by identifying factors and goals important to the DOE and its participants and then evaluating the SSABs and public affairs efforts for the ability to fulfill those goals (Bradbury and Branch 1999; Bradbury et al. 2003; Carnes et al. 1998; Weeks 2000). Carnes et al. (1998) documented attributes of successful participation based upon goals such as "the decision-making process should be seen as legitimate by citizens and should allow for full representation," as well as "DOE should understand public concerns." Bradbury and Branch (1999) found that six factors influenced the performance of SSABs, including the community's history of civic involvement, a diverse composition of board members, a commitment to consensus, processes such as team building and agenda setting, active communication with the community, and the involvement of DOE and EPA officials.

The analysis that follows differs from these past accounts of DOE public participation in an important way. I adopt a communication-based approach that thinks of public participation more broadly in terms of the ongoing dialogue, and I focus on the sense-making efforts of stakeholder groups that take place through these various formats. This approach examines the ways that Fernald stakeholder groups symbolically constructed the situation of nuclear cleanup and reveals the creative activities through which stakeholders discursively influenced evolutions in one another's perspectives over time.

PERSPECTIVES ON NUCLEAR WEAPONS CLEANUP

This study engages the struggle over nuclear symbols at one DOE site, the former Fernald nuclear weapons production facility, and does so at a point in time in which environmental cleanup became a central focus of the nuclear discourse among stakeholders. To understand the evolution of perspectives at Fernald, I examine sequential patterns of thinking and speaking expressed by DOE site officials, U.S. and Ohio EPA regulators, and members of the Fernald Residents for Environmental Safety and Health (FRESH) spanning from 1984 to 2003. I identify these patterns in stakeholder discourse found in public meeting transcripts, official documents, and interviews conducted as part of a local historical project.[2] For my purposes, a "perspective" represents what rhetorical theorist Kenneth Burke (1984) called *a frame of acceptance*, a system of meanings by which people label and interpret their experiences. The terms and labels people use to interpret and construct their experiences both give meaning to that experience and reflect particular understandings of the world. For Burke (1969), through the use of symbols people develop common understandings of the social world and form bridges or identifications with others.

Wertsch (1987) distinguishes between two dimensions of nuclear discourse that are helpful in understanding nuclear perspectives. The first, *scope of identification*, is the way in which a person sees his or her connection to others. It is the group a person identifies with, whether the family, nation, or humankind. The scope of identification can be either *particularistic*, viewing issues as a matter of "us versus them," or *universalistic*, viewing issues as "we're all in this together." Second is *a form of legitimation*, explained by Wertsch as "the processes of reflecting on, explaining, and defending decisions" (1987, 105). The form of legitimation can be either *de-contextualized*, using technical, abstract reasoning that transcends the particulars of a case, or *contextualized*, relying on situated instances and humanistic appeals, which Wertsch envisioned as primary ways of justifying arguments within the nuclear arms debate. Wertsch identified combinations of these categories. For example, a universal perspective of "we're all in this together" could be based either upon the contextualized thinking that people from all nations are subject to the same nuclear predicament or it could be based upon de-contextualized arguments such as official discourse of mutually assured destruction and the reality that a first strike means all humans are impacted. These categories provide a useful tool for exploring nuclear discourse in a post–Cold War period in which environmental and human health issues have become a focus.

THE EMERGENCE OF PUBLIC DISCUSSION AT THE FERNALD SITE

Following World War II the Atomic Energy Commission (hereafter, AEC) identified a 1,050-acre parcel of land located in the farming communities of rural southwestern Ohio to develop as part of the government's effort to expand nuclear weapons production. Originally called the Feed Materials Production Center, the Fernald site refined uranium ore and produced uranium metals from 1953 to 1989 as part of the U.S. nuclear weapons production complex. These uranium products were "fed" to other DOE sites to support defense programs (FDF 1998). Silverman (2000, 265) has noted that, "National security provisions, designed to safeguard atomic secrets, were also used to shield production facilities like Fernald from external environmental oversight and public accountability." This lack of oversight standards, coupled with the AEC's (and later DOE's) ability to self-regulate, meant that environmental and waste management concerns were largely ignored during the plant's thirty-year operating period. Consequently, in the mid-1980s, associated contaminants were found in nearby air, soil, and water. Of the total amount of waste produced over the lifetime of these operations, approximately 1 million pounds of uranium were released into the environment (FDF 1998, 1).

A series of events took place between 1984 and 1989 that shed light on the extent of environmental contamination and initiated public discussion in the surrounding communities. In 1984, a dust collector located in the plant's emission stacks malfunctioned and released nearly 300 pounds of enriched uranium oxide into the surrounding air (USDOE-FEMP 1995, Jan, 13). Following media coverage of this containment failure, concerned school officials hosted a public meeting to discuss the releases with DOE and its production contractor, National Lead of Ohio. For many citizens within this community this was the first time they were made aware of potential environmental damage, human health risks, and, for some, even the existence of the facility (Sheak and Cianciolo 1993). Also in 1984, the DOE disclosed that uranium had been previously discovered in 1981 in three off-site wells and two on-site wells. One of the off-site wells was the drinking water well of a nearby resident who later became president of the FRESH group (USDOE-FEMP 1995,

Jan). Local DOE officials held community meetings in late 1984 and early 1985 to take responsibility for producing this contamination. They were met with mistrust and skepticism from Fernald-area residents who, along with their elected representatives, began urging government agencies to investigate the extent of contamination at the site.

Reports of contaminated groundwater, fear of health effects, and anger that the DOE was not forthright with information about the site, evoked a response from a core of community members that has resulted in one of the most noted instances of civic involvement in the nuclear weapons complex (Sheak and Cianciolo 1993). Residents living nearby formed a grassroots organization in 1984 under the name Fernald Residents for Environmental Safety and Health (FRESH). The group emerged as a major voice for the community during the years following the public announcement of contamination, and for over two decades has been a driving force in decisions concerning the Fernald site (USDOE-FEMP 1995, Jan). Sheak and Cianciolo (1993, 116) note that since the mid-1980s FRESH has been accepted by the government as "a legitimate representative of the Fernald population."

Then in 1985, community members turned to litigation. A $300 million dollar class-action lawsuit was filed against DOE contractor National Lead of Ohio on behalf of 14,000 area residents living within five miles of the site. The suit was filed by several area residents, including the FRESH president, for emotional distress and decreases in property values caused by the site (USDOE 1995, 72). The lawsuit was settled four years later when the DOE paid $78 million in damages to Fernald-area residents including a medical-monitoring program for residents within a five-mile radius of the site that would help to determine potential health effects of production on individuals within the community.

Public scrutiny increased as the class action lawsuit progressed, and local and national media coverage as well as congressional investigations focused on Fernald. A DOE report noted that "150 reporters came to the facility over a ten-day period" (USDOE-FEMP 1995, Jan, 15). In 1988 and 1989, national magazines such as *Time, MACLEANS,* and *Good Housekeeping* reported the Fernald story as part of their coverage of the mounting charges of environmental safety concerns and neglect of human health by the DOE at many of its sites. For example, *Time* reporters Cramer et al. (1988, 62) wrote: "A bitter sense of betrayal, even among some defense-minded residents, has grown from the apparent aloofness of Washington officials to the perils that weapons production may pose to the health of innocent people living near the plants." This coverage emphasized classic narrative concerns of mystery and transgression: contractors operating without government oversight, unreported environmental releases, scientific uncertainty about the health effects of radiation, and the absence of a safe permanent disposal site for the wastes.

In the years immediately following the uranium leaks, the site came under the purview of both federal and state regulators. Ohio EPA inspected the Fernald site in 1984 and found that it was out of compliance with hazardous waste laws (Golightly 1993, 10). In 1986, the State of Ohio brought a lawsuit against the DOE and its contractor, National Lead of Ohio, for violating environmental laws. The U.S. Supreme Court ruled in 1992 that the State of Ohio could not assess penalties for past violations, but could issue injunctions to abate contamination or to begin a cleanup (Ohio EPA, OFFO 2001). The CERCLA process began at Fernald in 1986 when the DOE and U.S. EPA entered into a Federal Facility Compliance Agreement, followed by a series of interagency agreements that specified the relationship among these government agencies together with the legal milestones for Fernald site remediation (FFI 2002). Production was temporarily halted in 1989 and the Fernald site was placed on the National Priorities List for Superfund Cleanup. The site officially switched from weapons production to environmental remediation in 1991 at which time the name was changed to the Fernald Environmental Management Project (hereafter referred to as the Fernald site). The period of study and remedy selection took ten years, from 1986 to 1996. At the completion of the research for this chapter, the Fernald site was in the remedial action stage of CERCLA with completion of cleanup remedies planned for the end of 2006 (FFI 2002). Remediation activities at Fernald are expected to cost a total of $4.1 billion. Once remediation is complete and the Fernald site is removed from the National Priorities List, it will remain in federal hands. An on-site disposal facility and a restoration project for the Great Miami Aquifer beneath the site will require long-term stewardship in the form of both engineering and institutional controls such as public access restrictions, maintenance of the property, and monitoring tasks (USDOE 2001).

Fernald public discussion began with these revelations of contamination, public meetings, news reports, lawsuits, and increasing environmental regulation in the mid- to late-1980s. Since that time, a relatively formal and extensive public participation program has developed within Fernald-area communities as a result of a combination of grassroots efforts and institutional programs. The cleanup of the Fernald site includes mandated public participation activities specified for Superfund sites such as public notifications, comment periods, and public hearings. In addition to these legal requirements, Fernald site officials responded to the growing sentiment in the early 1990s that citizen engagement was essential by augmenting public participation with several innovations (Bradbury et al. 2003). Site officials have attempted to build awareness and trust within the community by providing greater and earlier access to information, involving site managers, and most notably by developing face-to-face communication and personal relationships with community members and opinion leaders in the area (USDOE-FN 1993). An important aspect of public participation at Fernald has been a SSAB organized in 1993 at a time when the DOE, Ohio and U.S. EPAs, FRESH, and area residents were beginning to collaborate and decisions were approaching that would define the extent of Fernald's remediation.

In the early to mid-1990s a transition took place in the relationship among stakeholders at Fernald which was tied to their changing views of public participation and how to develop cooperative ways of working toward site cleanup (Hamilton 2003b). DOE, EPA, and FRESH stakeholders had developed a common understanding of important elements of participation and their interaction with one another. For these stakeholder groups, a "good public process" should be based upon open communication, access to site management, early and continual participation in the day-to-day activities of the site, building relationships of trust and respect, and opportunities to learn and analyze the issues (Hamilton 2004). The Fernald SSAB and, to a lesser extent, regular public meetings brought together key stakeholder groups and structured their work through interactive forums with dialogue and group learning, identified above as the trend in public participation at that time. Through such efforts, stakeholders began to agree upon what constituted a good public participation process. These common understandings of process created a platform for the dialogue regarding environmental remediation of the site.

CONVERGENCE AND DIVERGENCE IN FERNALD'S PUBLIC DIALOGUE

In the ongoing public discussion at Fernald, two significant shifts have taken place in stakeholders' understanding of their relationships and the problems and solutions at hand. Though these shifts have been gradual, they can be captured through a series of defining moments that initiated a transition in understanding among community members, site officials, and regulators. In what follows, I examine these shifting symbolic constructions, and the associated discourses and practices by which these groups transformed their understandings of their relationships and nuclear cleanup issues.

Overcoming the Past

As Fernald stakeholders began orienting to one another and the task ahead, CERCLA requirements served as an organizing principle that structured the ways stakeholders understood solutions to the contamination and the tension among groups. The sense of betrayal and distrust among FRESH and area residents translated into an "us-versus-them" understanding of the situation. This framing was expressed in the following comments by a community member:

> [The DOE has] been dragged kicking and screaming to any sort of oversight . . .DOE has deceived us for many years and we're not likely to believe that they are willing to . . . be open and above-board about this. . . . This is very unbecoming of a federal agency that is supposed to be providing for our national defense. (US DOE-FMPC 1989, May 15, 33)

CERCLA provided part of the answer by involving regulators in ensuring a responsible cleanup. The other part of the answer lay in a community watchdog effort, reflecting a contextualized view that Fernald's solutions would require community input and oversight. Over a number of years, the FRESH group taught DOE their expectations as a watchdog group. FRESH emphasized that access to information was the foundation for developing trust between DOE and the community:

> When you [DOE] are asked a question, more often than not you cannot answer and promise you will get it. The answer many times is never given. . . . Still you refuse to notify people of possible or probable danger of contamination outside of the FMPC [the site]. . . . I feel that the residents can look forward to more exposure, more inaccurate information, and more well-founded fears. (US DOE-FMPC 1990, June 12, 15–18)

FRESH members recognized that, rather than blatantly withholding information from the public as in the past, DOE controlled information in new ways by restricting their participation through the release of untimely or inaccurate information or by failing to notify citizens about related incidents. In characterizing these practices, the FRESH president spelled out their implications for the group's relationship with DOE: "We have an open door policy here? And you sit up there and you wonder why we don't trust you and we don't believe you. That's exactly why" (USDOE-FMPC 1990, May 22, 131).

The FRESH group developed several strategies for turning this situation around. CERCLA requirements were contributing to the "us versus them" situation by restricting citizen input to comment periods and hearings at specific points in the process. A FRESH member later reflected on how the group overcame this challenge by gaining earlier access to needed information through draft documents:

> There again was the trust factor. They were afraid to let you see a draft document because you would say what's in it is gospel and once they realized we understand what "draft" means, you know, once we did some trial runs and they

found out we understood and weren't going to hold them to that, and that, we could work on stuff, it helped a lot. (FLHP 2000, November 30, 7)

FRESH essentially demonstrated that DOE could trust them with information before it was final or approved by the agencies. As a result, DOE began sharing information earlier and in draft form, affording FRESH the opportunity to offer their views before official review periods and hearings. Another strategy was to encourage DOE officials to be open with the community about scientific uncertainty: "The lie is they said it wasn't anything to be worried about. If they said [they] didn't know, that would be different" (US DOE-FMPC 1990, August 16, 103–112). FRESH constructed this as a situation in which they did not expect DOE to have all the answers, but they did expect them to be forthright. Through demands for timely and accurate information, candor regarding problems and uncertainties, and early and ongoing involvement in evolving decisions, FRESH set expectations for an open flow of information that would build trust between them. FRESH had created the possibility of overcoming the "us versus them" if DOE could engage their expectations for information and access to the process.

In contrast to FRESH, both government groups oriented to CERCLA requirements through discourse reflecting a "we're all in this together" perspective. An Ohio EPA regulator commented: "We've got to follow this process through. It's not going to be an easy one. We're going to have a lot of tough meetings . . . We're going to have to work together, and get our comments together, and come up with a way of cleaning up the site" (USEPA 1990, May 9, 69–70). This construction of the situation as one in which the stakeholders would "work together" and through the process set by Congress was grounded in a de-contextualized perspective that solutions would be found and protections would be ensured through a scientific and legal process. CERCLA solutions were a rallying point to shift attention away from their differences and to the political and technical complexities they would all face together in working through this process. It positions them to work on the same "side." DOE officials too understood their emerging commitments as following the prescribed legal process: ". . . and the commitment that you have from the U.S. government is that we are cleaning this material up and we are going through the process that has been established by the U.S. government, all of us in this room" (US DOE-FMPC 1990, August 16, 66–68).

As the various groups were defining their roles relative to the CERCLA process, an important discourse emerged in the talk of DOE officials which signaled the beginnings of the transition that would take place in stakeholder relationships. DOE officials working through CERCLA with the community described themselves as what I will call the "new guard" of agency officials committed to remediation of the site. This commitment was articulated by differentiating the past from present actions. References such as "the people that are in here now" and "you're talking to the RI/FS people, we're the people responsible for cleanup" (USDOE-FMPC 1990, May 22, 115) distinguished the old guard responsible for uranium refinement from the new guard of DOE officials responsible for remediation. This practice of differentiating the past from the present allowed them to talk about their current commitment without acknowledging past wrongs: ". . . we're not going to make excuses for past actions, but we're going to do all we can to try to prevent that from happening again" (US DOE-FMPC 1989, October 24, 143). In this manner, officials asserted that part of their commitment was to doing things differently than had been done in the past.

Separating the past from the present and those in charge of production from those responsible for remediation was a discursive practice that positioned these officials to work cooperatively with the community. It also was an attempt to persuade community members of their view that "we're all in this together." DOE officials portrayed themselves as experts providing the resources to study and solve the contamination problems faced by area communities. We can see the precursors of change for the relationships among stakeholders in the DOE's "new guard" discourse and FRESH's expectations for information and access.

Developing Points of Convergence

In the early to mid-1990s, these actors struggled to re-define the role of citizens in a government cleanup project and to assert various interpretations of the environmental and health risks of nuclear weapons production. Weapons production ended and the site mission officially changed to cleanup. The FRESH group demanded greater levels of involvement in decision-making and a SSAB known as the Fernald Citizens Task Force was convened and released recommendations regarding key issues such as cleanup levels, waste disposition, and future land use. "Records of Decision" finalizing the cleanup remedies for each of five areas of the site were reached under the CERCLA law. Public discussion turned to issues such as the extent to which the site would be cleaned, where the waste would be permanently disposed, and the specific methods by which cleanup would proceed.

Stakeholder perspectives converged during this time in an emerging view of themselves as a "we" on the local level addressing these issues together. This convergence was achieved in part through descriptions of a joint effort and a local partnership as well as a practice of differentiating site from federal level DOE. An important shift was taking place in which solutions to Fernald's problems were increasingly seen as a collaborative effort on the local level among the DOE, the EPAs, and residents including the FRESH group. While perspectives were converging in terms of some issues, it is important to note that each group maintained perspectives consistent with early public discussion at the same time. This first shift was accomplished through discourses that gradually ushered in new understandings of the relationships among the groups and a sense of cooperation in addressing these issues. Below I examine two defining moments in this gradual transition.

Securing the Budget through an Accelerated Plan

As key decision points in the CERCLA process were approaching, DOE Headquarters announced environmental cleanup budget cuts that could potentially impact Fernald remediation work. These cuts could have extended the total length of time required for remediation and cost more money overall. An Ohio EPA regulator described the situation as one requiring all the stakeholders to work together to secure the necessary funding:

> The real work is doing the cleanup, and that's where we're all going to have to work together as partners as well because it's going to be very difficult. Already you see the federal government losing some of its focus on the cleanup of DOE sites. Funding is being cut, and these groups here, everybody, is going to have to work very hard to keep the funding, keep the attention focused on Fernald and other DOE sites in the complex to make sure that the cleanup is taking priority. (US DOE-FEMP 1993, October 21, 25)

The construction of the problem as one in which the federal government was losing focus served to differentiate DOE officials at the site from DOE officials at Headquarters. The problem was not articulated as the DOE per se, but the DOE at the federal level. From the regulator perspective, site-level officials were working alongside regulators and citizens to ensure an adequate budget. Using a strategy described by Cheney (1983) as *identification through antithesis*, the EPA official had defined a common enemy in pointing to Headquarters as the source of the problem. Such a dissociation of local from federal DOE was an attempt to connect the local actors by allowing them to identify with one another and against Headquarters as a common enemy. As we saw with the new guard discourse, the practice of differentiating the "good" from the "bad" DOE officials was a mechanism by which a shift in relationships was being accomplished.

An accelerated or "ten year plan" was proposed by Fernald site officials out of concern that it would take twenty-five years to complete the remediation of Fernald because of the projected budget cuts. Rather than reducing the budget, an accelerated schedule would require investing more money in the near term, but it would enable a quicker cleanup, and would reduce the costs from $5.7 billion to $2.9 billion (FCTF 1995). An important part of this effort was the FRESH group's willingness to work alongside site officials in urging DOE Headquarters to sustain Fernald's budget levels. A FRESH member explained:

> I've been to headquarters once . . . to talk to Mr. Grumbly [DOE Assistant Secretary for Environmental Management] about budget stuff. . . . We've been talking to everybody and their brother that we can talk to about this, around here and in Washington. (USDOE-FEMP 1993, June 22, 77–8)

Eventually, DOE Headquarters adopted this accelerated cleanup plan for Fernald. An Ohio EPA regulator pointed to the joint effort among stakeholders as contributing to the reception of this idea:

> Through everybody's hard work we see a lot of support for the ten-year plan out there both in Congress and in DOE headquarters, and I think we all have earned the recognition of those people for this new and innovative approach to getting the site cleaned up. (US DOE-FEMP 1995, August 8, 109–110)

In this way, regulators portrayed "everybody's hard work" as contributing to the acceptance of an accelerated cleanup plan and an adequate level of funding. Likewise, FRESH members spoke about a joint effort: "We've done a lot, we've accomplished a lot. We've worked extremely hard on the budget. . . . But I think it did take a nice, good, joint, hard effort to get where we are at now" (USDOE-FEMP 1995, August 8, 111–112). An important transition was taking place in stakeholder understandings through such descriptions of a joint effort and the construction of an "us" at the local level seen as a way to overcome the obstacles posed by a "them" of the federal government losing sight of cleanup as a priority.

Finding a Balanced Approach to Waste Disposal

The next moment that captures the emerging convergence involves the three groups in a shared understanding that the solution to Fernald's waste disposal was a "balanced approach." As the Records of Decision were approaching, the DOE, regulators, and area residents began grappling with options of where to permanently dispose of Fernald's waste, including off-site at a DOE facility in Nevada, off-site at a commercial facility in Utah, and/or on-site in an engineered facility that would be built at Fernald. The DOE's preferred alternative was on-site disposal of a high volume of low-level waste and off-site disposal of a lower volume of highly contaminated waste. The three groups' construction of this solution as a balanced approach reflects a shared "we're all in this together" perspective in which the "we" was broadly defined as Fernald-area communities, receptor communities, and those along the transportation routes. One DOE official indicated: "There's also a community in Utah and communities in every state through which material must pass for off-site disposal" (USDOE-FEMP 1994, November 8, 20).

Because Fernald decision-making could impact neighboring communities at other sites, FRESH emphasized the need to understand waste disposal issues in relation to the larger nuclear weapons complex. According to the FRESH president, cleanup decisions required an understanding of ". . . the national picture, a look at everywhere, not just here, and that's what people are going to need to make these tough decisions on land use and waste and how clean is clean and all those kind of issues" (USDOE-FEMP 1993, October 21, 34). FRESH members expressed reluctant support of the DOE's balanced approach, describing that, although they did not want the waste disposed on-site, they were willing to accept it because of the need for a safe and equitable disposal location. One FRESH member described:

> I live in Crosby Township, less than a mile from the site, and I am willing to accept the preferred alternative because there are a lot of other people in this country who are dealing with this same issue, and they don't want this stuff in their backyard either, and if we can get the worst of this stuff out of here, I think the least we can do is be responsible for what we can safely keep here. (USDOE-FEMP 1994, November 8, 49)

FRESH member references to needing to understand the national picture and reluctance to put the waste in someone else's backyard reflect development of a universalistic "we're all in this together" perspective. The FRESH, DOE, and EPA groups had constructed a balanced solution as the best solution in ways that reflected a shared contextualized perspective, grounded in the situated circumstances of Fernald-area and receptor communities.

The movement toward shared perspectives for understanding both the budget resolution and waste disposal was accomplished in part through the construction of themselves as a "we" of site-level actors who would find local solutions to funding hurdles posed by the federal government and to a safe and equitable waste disposal location. These shared meanings for the accelerated plan and the balanced approach were central to stakeholders' ability to collaborate despite lingering issues of distrust.

This initial shift in perspectives to convergence was influenced by several situational factors as well. It happened near the end of a tremendous time of cultural change that took place in the peoples and institutions of the DOE.

The agency was learning to involve the public in an arena that had operated without scrutiny or oversight for decades. The Clinton administration's initiatives to "reinvent government" and create accountability influenced a changing politics and new tone within DOE. Secretary of Energy Hazel O'Leary began an "openness initiative" to de-classify thousands of formerly classified documents, creating a new stance toward acknowledging the public and their right to information. This change was effected at Fernald by DOE site managers who represented the new guard of the DOE focused on environmental remediation and by public affairs officers who designed public participation around one-on-one communication, management involvement, small group formats, and a SSAB.

The shift to convergence was also influenced by changes within the FRESH organization. The leaders of FRESH in particular had spent several years educating themselves in terms of the technical and legal knowledge they needed to contribute to the decision-making process. FRESH leaders adopted a strategy of participating both inside and outside the process, initially gaining voice through a lawsuit and heated discussions at public meetings, and later by influencing the discussion from within the federally-charted Fernald SSAB. Significantly, the group framed their concerns in terms of public and environmental health issues rather than antinuclear sentiment (i.e., opposition to weapons production per se), thus removing a historical obstacle for collaboration between nuclear citizens and officials, and allowing them to focus their shared attention on the DOE's responsibility for the cleanup.[3]

Changing legal requirements also influenced the initial convergence of stakeholder perspectives. Conflict among stakeholders emerged when the site was producing weapons grade materials and was still self-regulating according to Atomic Energy Act provisions. The Fernald site's inclusion on the National Priorities List for Superfund Cleanup in 1989 and later the Federal Facilities Compliance Act of 1992 created obligations for the DOE to comply with environmental laws. These changes in the application of environmental laws to federal facilities provided legal requirements for the DOE to involve the public in decision-making.

Finally, FRESH, EPA, and DOE stakeholders participated in the Fernald SSAB. The issues that prompted the convergence examined here were issues deliberated by the board and addressed through its initial effort and recommendations. In this way, the SSAB served as a key forum for bringing the various stakeholders together and structuring group learning, analysis, and dialogue among them. The SSAB influenced the emerging convergence by creating opportunities for stakeholders to develop and express their own perspectives, to explore others' perspectives, and to create and transform the associations among them. In this way, the persons, institutions, and clear legal drivers that dominated the early years of Fernald public discussion developed a context conducive to interaction and influence among diverse perspectives that led to a certain degree of convergence during the early to mid-1990s.

The Re-emergence of Divergence

As the Fernald site moved into the CERCLA phase in which cleanup remedies are designed and implemented, decision-making turned from how the site should be cleaned to how the land should be used post-remediation. In the late 1990s and early 2000s, a public process began of identifying the future uses of the Fernald site and a mechanism for long-term stewardship. During these stewardship discussions stakeholders sensed that the DOE had begun retrenching on its commitments to support future public uses of the Fernald land. An interesting evolution took place in which the FRESH and Ohio EPA groups assessed problems and solutions concerning public use and long-term stewardship in ways that revealed the re-emergence of an "us-versus-them" perspective set against local DOE officials.[4] These issues were constructed by these groups as a matter of ensuring that the DOE would honor its agreements, leave the community with a viable resource, and provide the stewardship to maintain it. Similarly to the initial shift, this transition was partial and points of convergence continued especially in relation to views of the public participation process. Through discourses that questioned the understandings and cooperation that had developed earlier, divergent perspectives began to resurface. These changes can be captured in stakeholder discourse during two key moments that signaled a transition was underway.

Turning a Negative into a Positive through Public Use

The first moment involves public discussion regarding the future uses of the Fernald land. FRESH members and Ohio EPA regulators constructed future use issues in terms of the DOE providing a benefit to area communities in ways that suggest a change was in progress in their understanding of the situation. The FRESH president explained: "[I]t is time to begin shaping a long-term vision for the future of the Fernald site that turns a negative into a positive and creates something of lasting value for the surrounding communities" (USDOE-FN 1999, June, comment 1). The phrase "turns a negative into a positive" captures the sense among FRESH, Ohio EPA, and other community members that land-use decisions presented an opportunity for the DOE to make amends for the harms caused to this community. For regulators, overcoming the Fernald stigma would involve transitioning the site from a Superfund cleanup project to a resource for the community: "That's our challenge is to reach out beyond the folks that know Fernald as the nightmare, the contamination scare, the place that contaminated me, to the new generation that sees it as a resource" (FLHP 2000, November 3, 16).

FRESH, Ohio EPA, and other area residents articulated several land uses that would represent such a resource, including green space, an educational center, and the reburial of Native American remains. The FRESH president commented that an educational center would "create something of national significance and offer broader and longer-lasting value to the community" (USDOE-FN 1999, June, comment 1). Regulators emphasized that a center could act as an institutional control through educating citizens about what is required to continue protecting the environment and surrounding communities from residual contamination. It would be a positive use of the land that would help to overcome a view of Fernald as contaminated or spoiled.

Within this perspective, an "us" comprised of people who would use the land after the closure of the site were seeking decisions from a "them" of the DOE responsible for harming the land and creating a stigma for Fernald-area communities. FRESH and Ohio EPA regulators constructed this as an opportunity for the DOE to "give back" and to make amends for the harms they had caused in these communities by leaving them with such a resource. This discourse of "turning a negative into a positive" represents the beginnings of a shift away from the earlier partnership in which DOE site officials were part of the solution.

Retrenching on Stewardship Commitments

The shift to divergent perspectives intensified as FRESH and Ohio EPA regulators reacted to the DOE's "Comprehensive Stewardship Plan" as a situation in which the DOE was retrenching on its commitments by promoting a limited definition of its responsibilities for long-term stewardship at the site.[5] The DOE had engaged the public in a planning process for the future uses of the site, but was backing away from the stewardship needed to support the uses identified by the community as creating a positive resource. As the FRESH president described: "DOE is beginning to pull back and say they're not going to follow through with some of the promises we've agreed to over the years" (McAllister 2003). In its stewardship plan, the DOE committed to monitoring and maintenance of the on-site disposal facility and other certified areas of the site, but not of the restoration projects or other "public use amenities" such as walking trails and an educational center. Regulators too understood the plan as inconsistent with previous commitments made by the DOE:

> That document [Master Plan for Public Use] was the culmination of significant public involvement and was designed to be the equivalent of NEPA documentation for the decision to develop the site . . . [which] provided a commitment from DOE to the public at Fernald. The decision by DOE to disregard these documents concerns the agency and should provide great concern to stakeholders regarding the reliability of DOE to follow through with commitments it makes. (Ohio EPA 2003, January 17)

In this way, FRESH, Ohio EPA, and other community members understood public uses identified in conjunction with the DOE as part of a "good faith" agreement with the community. One FRESH member described: "We took the higher road and kept 75 to 80 percent [of the] waste on site. We really have done our part and they should have to do theirs" (McAllister 2003). The phrase "we've done our part" was a reference to the balanced approach and FRESH's acceptance of a low-level waste disposal facility on the Fernald property. For the FRESH and regulator groups, what the DOE needed to give in return was the stewardship needed to support the community vision of the site. When the stewardship plan was released, however, FRESH members saw it as addressing only minimum activities needed to fulfill legal requirements. A FRESH leader commented in a local newspaper:

> They're only doing the minimum wherever possible. They'll only come in every five years to monitor the disposal cell, and when you start leaving a site looking like that—with a fence all around it and without amenities and that has the reputation of being radioactive—it will economically destroy the community if it's not made a valuable asset. (McAllister 2003)

Ohio EPA regulators saw the current DOE interpretation as focusing on regulatory requirements for monitoring and maintenance without necessarily providing a benefit to the community. One regulator pointed to DOE Headquarters for explanations of why Fernald officials would drastically alter agreements with the community saying, "DOE unilaterally changed the document based upon the latest 'direction' from HQ" (Ohio EPA 2003, January 17). This construction of the problem is reminiscent of the budget issue in which the problem was understood as a matter of federal level DOE posing hurdles to the cleanup funding. Here, federal DOE officials were seen as directing site officials to focus on minimum requirements for long-term stewardship. Rather than the local interpretation of stewardship, site officials followed DOE Headquarters' interpretation, regressing to the status of a "them" working at cross purposes with regulators and community members. This construction cast the "latest direction" from Headquarters as causing a breach in cooperation on the local level that had sustained decision making for a number of years. Reconfiguring the local DOE with Headquarters as part of the common enemy was a key practice that solidified the divergence in perspectives.

Much of the work that had been done historically to develop a "we" of local actors was undone by the stewardship issue. An "us" of community members and regulators were struggling against the DOE to actualize the end-state that had emerged through a public process. A FRESH leader commented, "It's just like we're starting all over again" (McAllister 2003). For FRESH members, this particularistic perspective was a return to earlier ways of interpreting issues before they developed a certain level of trust in the DOE. However, the Ohio EPA's strong sense that the DOE was "unilaterally changing directions" and "disregarding documents" reflects the emergence of an "us versus them" perspective in seeing the DOE as a hurdle to achieving the community vision for the future use of Fernald.

Just as the shift to convergence had been influenced by a tone and priorities set within the DOE as well as changing legal requirements, so too was the shift to divergence. CERCLA requirements for public participation were considerably less during the implementation phase of the process. Beyond preparation of an Institutional Control Plan, there are no clear legal drivers for the DOE to commit to stewardship activities (Wells and Spitz 2003). The lack of a legal driver, such as a Record of Decision, meant that the DOE was fulfilling minimal legal obligations rather than supporting aspects of the community vision of the future of Fernald. FRESH and the regulators understood this inability to commit without a legal obligation as grounds for distrust. The DOE stance of doing the minimum was in turn influenced by Headquarters and a new tone and priorities within the DOE. As was true during the initial shift in perspective, this tone was set in part by the post 9/11–era Bush administration, and reflected the shifting of national priorities to pressing issues such as homeland security. The Fernald SSAB that had been instrumental in cleanup decisions also coordinated a public process and provided recommendations on future use decisions, but this effort garnered little agency attention (FCAB 2002). Therefore, decreasing legal requirements and changing politics influenced this re-emergence of an "us versus them" perspective on stewardship issues.

FINDINGS AND CONCLUSIONS

What has the Fernald case taught us about the communicative ability of various interests in nuclear debate to inform and influence one another's perspectives? Over time the development of public participation at Fernald provided the opportunity for competing nuclear interests to interact in ways not possible during the Cold War. This long-term interaction involved assertions, contestations, and refinements of nuclear meanings through public forums that encouraged dialogue and group learning, and that allowed stakeholders to develop relationships and influence one another in ways that facilitated change in aspects of their perspectives. A productive tension developed among the DOE, EPA, and FRESH groups in the ways in which they constructed the problems and solutions of environmental cleanup. This productive tension points to two findings that are significant to the study of nuclear public participation.

First, this analysis has revealed an evolving pattern of convergence and divergence of perspectives when discourse is considered over time. The convergence that gradually emerged was achieved in great part through the construction of a "we" of local or site-level actors who would need to collaborate to work through the CERCLA process and clean up the site. For the budget issue this was articulated by all the groups as a matter of local actors working together as an "us" against federal actors to secure the cleanup budget and an accelerated schedule, and the waste disposal issue as one in which "we're all in this together" in striving for a balanced solution to the safe and equitable disposal of the wastes. This perspective was fostered through strategies such as separating the past from the present and the old guard from the new guard of the DOE, as well as strategies that defined a common enemy for the groups in viewing the federal level DOE as a hurdle to sufficient funding while describing the local actors as working through a joint effort. It was also fostered through the development of interpersonal relationships and a degree of trust and respect through participation techniques that reached beyond legal mandates and structured a dialogue conducive to influence and change in perspectives. The divergence that resurfaced was similarly influenced by a shifting construction of the "us" and "them," in which local officials were seen as retreating from their associations with local actors in terms of future use and stewardship issues. Importantly, remnants of distrust and skepticism lingered even as convergence built among stakeholders and, likewise, remnants of cooperation continued even as stakeholders questioned their relationships.

Second, despite the re-emergence of divergent perspectives, stakeholders were able to move forward and continue in dialogue. The Fernald case has shown that dialogue can occur even as divergence continues alongside emerging points of convergence and, importantly, dialogue can continue in the face of a shift to divergence. Dialogue can be seen as a process in which convergence and divergence exist simultaneously or alternately, rather than a process that moves in one direction toward resolution such as consensus on a course of action.

The three Fernald groups were able to sustain dialogue and productive relationships even as conflict erupted again in large part because of their shared belief of what constituted good public participation. As convergence was developing, the formats for public participation and the ways in which the groups interacted greatly expanded. The DOE began using small group sessions during the public meetings and began offering roundtables and workshops on issues of particular concern to community members. The site began implementing a philosophy of participation based on one-on-one relationships with community leaders and an involved management team. During small group settings, area residents were able to interact with site officials and regulators, asking detailed questions and exploring the issues in informal, conversational formats. The quality of discourse changed as small group forums created opportunities for competing interests to mingle and interact. Stakeholders came to understand each other's perspective, its bearing on a particular issue, and the relations among perspectives. As this understanding gradually grew, stakeholders began to jointly define problems and evaluate solutions.

Many of these productive participation practices continued even as divergence was resurfacing. Significantly, the network of interpersonal contacts continued to be a sustaining force in the interactions among officials and FRESH members. DOE, EPA, and FRESH groups continued to operate out of the principle of early and ongoing involvement that was central to the partnership that had formed. When conflict re-emerged, therefore, it was among stakeholders who knew and understood each other's positions on the issues. They continued to share an underlying perspective on public participation as the basis for discussing competing interpretations of stewardship. The continued dialogue then resulted not from any given participation mechanism or set of criteria for a good process, but ultimately from the sustained nature of the relationship among stakeholder groups.

These findings have implications for scholars and critics of public participation, and particularly those looking at these processes within the nuclear weapons complex. First, this study demonstrates that dialogue can happen without fulfilling normative conditions for democratic participation and communication. Most public participation studies have not examined long-term participation efforts and have not identified such an evolution. The literature on democratic criteria and undistorted communication tends to assume that "good process" will lead to "good outcomes" such as collaborative relations among participants, meaning that is openly formed, or consensus on politically legitimate decisions. Considering the evolution of perspectives over time in the Fernald case has shown that a more democratic form of participation did significantly influence movement toward shared understandings and productive relations among diverse interests. Notably, however, many of these participation processes remained in place even as conflict re-emerged and perspectives diverged, thus challenging our conception of a "good outcome" as one that *resolves* conflicting viewpoints and opening a space to consider a good outcome to be simply an ongoing conversation among stakeholders who understand and respect one another's points of view.

Likewise, past studies of the DOE's public participation programs in particular have defined and measured the success or effectiveness of such efforts largely through process and structural elements (Bradbury and Branch 1999; Bradbury et al. 2003; Carnes et al. 1998). At the Fernald site, these studies have documented success in the work of the SSAB (Applegate 1998; Bradbury et al. 2003; Williams 2002) or the FRESH group (Sheak and Cianciolo 1993). Success is seen in terms of organizing and becoming an accepted voice for the community, achieving concessions from the DOE, impacting DOE decisions, or achieving consensus recommendations. A key factor overlooked in this literature is the development of a participation structure that accommodates difference and conflict as a natural part of dialoging. Some have begun to question consensus as the goal of the deliberations and interactions of the DOE's SSABs (Weeks 2000). This study encourages and extends such discussions by revealing the ability of dialogue and public processes to continue despite enduring and emerging differences, and it challenges predominant definitions of success and effectiveness in this literature to include process elements that anticipate and accommodate conflict.

Second, this study has methodological implications for the study of nuclear discourse. The approach adopted in this study of tracking the evolution of understandings over time allows us to see transformations in perspective and the practices by which they are accomplished throughout sustained participation. In this chapter, stakeholder perspectives were identified using Wertsch's (1987) categories, a method that poses both benefits and limitations for the analysis. Using these categories gives us a consistent way to trace these evolutions over time. The dichotomy between an "us versus them" and "we're all in this together" perspective was strong in Fernald stakeholder discourse, providing a clear way to identify changes in the ways the groups envisioned their relationships. This approach, however, also limits our access to aspects of stakeholder perspectives pertaining to these categories and it offers little by way of analyzing the process by which the shifts in perspective took place.

In this respect my analysis employs a communication approach much like that of Mehan, Nathanson, and Skelly (1990) who analyzed late Cold War shifts in nuclear discourse, signaled by the Reagan Administration's representation of nuclear weapons as a mechanism to win war rather than prevent war. These cumulative shifts eventually "unraveled" many Cold War conventions centered on deterrence policy. The shift to cooperation at Fernald has been documented in other studies as resulting from the work of the SSAB and from a public affairs emphasis on management involvement and face-to-face communication (Bradbury et al. 2003; Williams 2002). Alternately, a communication approach allows us to focus on stakeholder discourse as a form of strategic action by which the actors *achieved* shifts in perspectives. This focus yields a view of the agency of change not afforded by sociological approaches that capture sense-making at a discrete point in time or by public participation analyses looking solely at structure or process elements. The approach adopted in this study allows us to capture not only frames for interpretation and how they change overtime, but also the discourses through which changes are effected. Through shifting constructions of the "us" and "them," stakeholder discourse ushered in views of the local actors as working cooperatively, and later at cross purposes. The shifts captured in this study were largely rooted in changing views of the relationships among stakeholders, Wertsch's category of "scope of identification." Changes in the ways they constructed associations and dissociations among groups in turn impacted how they viewed the problems and solutions at hand, suggesting the importance of this dimension in terms of creating a productive tension among groups, such as that seen at Fernald.

Across the complex, many sites and their neighboring communities are still grappling with a continued production mission and/or decisions about appropriate remediation and ways to create productive collaborations—and may never achieve the sustained participation seen at Fernald. It is clear that the challenge ahead for DOE public participation is how to sustain these processes throughout the period of long-term stewardship of these sites. This study offers important insights about promoting and sustaining dialogues through development of a productive tension among groups that are especially relevant for these sites. Fernald stakeholders have achieved some unique successes, but chief among them was the ability to continue in dialogue despite both continuing and rising conflicts. This was possible because they had achieved a common view of important elements of public participation, and it resulted not from a given mechanism but from the development of relationships over time.

NOTES

1. The author wishes to thank Stephen Depoe and Bryan Taylor for many critical insights in the development of this essay.

2. The Fernald Living History Project is a community-based effort including a local historical society, community leaders, the Center for Environmental Communica-

tion Studies at the University of Cincinnati, the DOE and its contractor Fluor Daniel Fernald, and the Ohio EPA. Since its inception in October 1997, project partners have conducted over one hundred interviews to create a historical archive of personal narratives of community and worker experiences with the site. The author has been involved in this project and has served as interviewer for twelve of these interviews.

3. Ratliff (1997a) similarly found that the activist group, the Hanford Education Action League, took a position on environmental, health, and safety concerns, but not on nuclear weapons.

4. A majority of the regulator discourse studied during this time period represents Ohio EPA regulators who participated more actively in land use, public use, and stewardship discussions as one of several Natural Resource Trustees charged with settling the claim for natural resource damages filed by the State of Ohio in 1986.

5. An evaluation study released in 2003 identified related concerns among stakeholders throughout the nuclear weapons complex that: "Although there was no formal change in public participation policy . . . DOE-EM Headquarters was sending a variety of signals indicating that they were placing a lower priority on consultation with community stakeholders" (Bradbury et al. 2003, vii) and "the declining influence of site personnel in decision-making" (Bradbury et al. 2003, 32).

BIBLIOGRAPHY

Applegate, John. 1998. Beyond the usual suspects: The use of citizens advisory boards in environmental decision making. *Indiana Law Journal* 73: 903.

Bradbury, Judith, and Kristi Branch. 1999, February. *An evaluation of the effectiveness of local site-specific advisory boards for U.S. Department of Energy environmental restoration programs. Report PNNL-12139.* Washington, D.C.: Pacific Northwest National Laboratory.

Bradbury, Judith, Kristi Branch, and Elizabeth Malone. 2003, February. *An evaluation of DOE-EM public participation programs. Report PNNL-14200.* Washington, D.C.: Pacific Northwest National Laboratory.

Burke, Kenneth. 1969. *A rhetoric of motives.* Los Angeles: University of California Press.

———. 1984. *Attitudes toward history*, 3d ed. Los Angeles: University of California Press.

Cable, Sherry, Thomas Shriver, and Donald Hastings. 1999. The silenced majority: Quiescence and government social control on the Oak Ridge Nuclear Reservation. *Research in Social Problems and Public Policy* 7: 59–81.

Carnes, S. A., M. Schweitzer, E. B. Peelle, A. K. Wolfe, and J. F. Munro. 1998. Measuring the success of public participation on environmental restoration and waste management activities in the U.S. Department of Energy. *Technology in Society* 20: 385–406.

Cheney, George. 1983. The rhetoric of identification and the study of organizational communication. *Quarterly Journal of Speech* 69: 143–159.

Chess, Caron, and Kristen Purcell. 1999. Public participation and the environment: Do we know what works? *Environmental Science & Technology* 33: 2685–2692.

Cramer, Jerome, B. Russell Leavitt, and J. Madeleine Nash. 1988. They lied to us: Unsafe, aging U.S. weapons plants are stirring fear and disillusion. *Time*, 31 October: 61–65.

Crowfoot, James E., and Julia M. Wondolleck. 1990. *Environmental disputes: Community involvement in conflict resolution.* Washington, D.C.: Island Press.

Daniels, Steven, and Gregg Walker. 1996. Collaborative learning: Improving public deliberation in ecosystem-based management. *Environmental Impact Assessment Review* 16: 71–102.

Duffield, Jennifer, and Stephen Depoe. 1997. Lessons from Fernald: Reversing NIMBYism through democratic decision-making. *Risk Policy Report* 3: 31–34.

Fehner, Terrence R., and Francis G. Gosling. 1996. Coming in from the cold: Regulating U.S. Department of Energy nuclear facilities, 1942–1996. *Environmental History* 1: 6–32.

Fernald Citizens Advisory Board (FCAB). 2002, December. *Memo from FCAB Chair and Stewardship Committee Chair to DOE Fernald concerning the comprehensive stewardship plan for the Fernald Environmental Management Project.* Fernald, Ohio.

Fernald Citizens Task Force (FCTF). 1995, July. *Fernald Citizens Task Force: Recommendations on remediation levels, waste disposition, priorities, and future use.* Fernald, Ohio.

Fernald Living History Project (FLHP). 2000, November 3. *Interview with Thomas Schneider.* Fernald, Ohio: Fernald Living History, Inc. Available from the Public Environmental Information Center 513-648-3000.

Fernald Living History Project (FLHP). 2000, November 30. *Interview with Pam Dunn.* Fernald, Ohio: Fernald Living History, Inc. Available from the Public Environmental Information Center, 513-648-3000.

Fiorino, Daniel J. 1989. Technical and democratic values in risk analysis. *Risk Analysis* 9: 293–299.

———. 1990. Citizen participation and environmental risk: A survey of institutional mechanisms. *Science, Technology, & Human Values* 15: 226–243.

Fluor Daniel Fernald (FDF). 1998, June. *1997 Integrated site environmental report for the Fernald Environmental Management Project* (51350-RP-0001). Fernald, Ohio: U.S. Department of Energy.

Fluor Fernald, Inc. (FFI). 2002, June. *2001 Site environmental report for the Fernald Environmental Management Project* (51350-RP-0019). Fernald, Ohio: U.S. Department of Energy.

Freer, Brian. 1994. Atomic pioneers and environmental legacy at the Hanford Site. *Canadian Review of Sociology & Anthropology* 31: 305–324.

Golightly, Eric J. 1993. *Site history of the Fernald Environmental Management Project.* Washington, D.C.: U.S. DOE, Office of Environmental Restoration & Waste Management, History Division.

Gusterson, Hugh. 1996. *Nuclear rites: A weapons laboratory at the end of the Cold War.* Berkeley: University of California Press.

Habermas, Jurgen. 1970. Towards a theory of communicative competence. *Inquiry* 13: 363–372.

———. 1984. *Theory of communicative action, vol. 1: Reason and the rationalization of society.* Trans. Thomas McCarthy. Boston, MA: Beacon Press.

———. 1987. *Theory of communicative action, vol. 2: System and lifeworld.* Trans. Thomas McCarthy. Boston: Beacon Press.

Hamilton, Jennifer D. 2003a. Exploring technical and cultural appeals in strategic risk communication: The Fernald radium case. *Risk Analysis* 23: 291–302.

———. 2003b. Sustained public participation and nuclear weapons cleanup: The evolution of stakeholder perspectives at the Fernald nuclear weapons site. Unpublished doctoral dissertation, University of Cincinnati.

———. 2004. Competing and converging values of public participation: A case study of participant views in Department of Energy nuclear weapons cleanup. Pp. 59–81 in *Communication and public participation in environmental decision making*, ed. Stephen Depoe, John Delicath, and Marie-France A. Elsenbeer. Albany: SUNY Press.

Ingham, Zita. 1996. Landscape, drama, and dissensus: The rhetorical education of Red Lodge, Montana. Pp. 195–212 in *Green culture: Environmental rhetoric in contemporary America*, ed. Carl G. Herndl and Stuart C. Brown. Madison: University of Wisconsin Press.

Juanillo, Napoleon K., and Clifford W. Scherer. 1995. Attaining a state of informed judgments: Toward a dialectical discourse on risk. Pp. 278–299 in *Communication yearbook 18*, ed. Brant Burleson. Thousand Oaks, Calif.: Sage Publications.

Kaplan, Louise. 2000. Public participation in nuclear facility decisions: Lessons from Hanford. Pp. 67–83 in *Science, technology, and democracy*, ed. Daniel L. Kleinman. Albany: SUNY Press.

Katz, Steven B., and Carolyn R. Miller. 1996. The low-level radioactive waste siting controversy in North Carolina: Toward a rhetorical model of risk communication. Pp. 111–140 in *Green culture: Environmental rhetoric in contemporary America*, ed. Carl G. Herndl and Stuart C. Brown. Madison: University of Wisconsin Press.

Keystone Center. 1996, April. *Consensus, principles, and recommendations for improving federal facilities cleanup. Final report of the Federal Facilities Environmental Restoration Dialogue Committee.* Washington, D.C.: U.S. Government Printing Office.

Killingsworth, M. Jimmie, and Jacqueline S. Palmer. 1992. *Ecospeak: Rhetoric and environmental politics in America.* Carbondale: Southern Illinois University Press.

Kinsella, William J. 2001. Nuclear boundaries: Material and discursive containment at the Hanford nuclear reservation. *Science as Culture* 10: 164–194.

Laird, Frank N. 1993. Participatory analysis, democracy, and technological decision making. *Science, Technology, & Human Values* 18: 341–361.

McAllister, Kristin. 2003. Residents: DOE not doing its part. Federal agency accused of backing out of Fernald responsibilities. *The Hamilton Journal News,* 23 January: p. C–3.

Mehan, Hugh, Charles E. Nathanson, and James M. Skelly. 1990. Nuclear discourse in the 1980s: The unraveling conventions of the Cold War. *Discourse and Society* 1: 133–165.

Mouffe, Chantal. 1999. Deliberative democracy or agonistic pluralism. *Social Research* 66: 745–758.

Ohio Environmental Protection Agency (Ohio EPA). 2003, January 17. *Letter from Thomas A. Schneider, Fernald Project Manager for Ohio EPA to Johnny Reising US DOE, Fernald Area Office concerning comments on the US DOE's January 2003 Comprehensive Stewardship Plan for the Fernald Environmental Management Project.*

Ohio Environmental Protection Agency, Office of Federal Facilities Oversight (Ohio EPA, OFFO). 2001, January. *Fernald Chronology*, offo2.epa.state.oh.us/Fernald/FernaldInfo/chronology.htm (accessed January 24, 2001)

Prosise, Theodore O. 1998. The collective memory of the atomic bombings misrecognized as objective history: The case of the public opposition to the National Air and Space Museum's atom bomb exhibit. *Western Journal of Communication* 62: 316–347.

Ratliff, Jeanne N. 1997a. Improving environmental advocacy: How the Hanford Education Action League challenged the Department of Energy. *Journal of the Northwest Communication Association* 25: 42–59.

——. 1997b. The politics of nuclear waste: An analysis of a public hearing on the proposed Yucca Mountain Nuclear Waste Repository. *Communication Studies* 48: 359–380.

Renn, Ortwin, Thomas Webler, and Peter Wiedemann, eds. 1995. *Fairness and competence in citizen participation: Evaluating models for environmental discourse.* Boston: Kluwer.

Santos, Susan L., and Caron Chess. 2003. Evaluating Citizen Advisory Boards: The importance of theory and participant-based criteria and practical implications. *Risk Analysis* 23: 269–280.

Sheak, Robert J., and Patricia Cianciolo. 1993. Notes on nuclear weapons plants and their neighbors: The case of Fernald. *Research in Social Problems and Public Policy* 5: 97–122.

Silverman, Joshua M. 2000. No immediate risk: Environmental safety in nuclear weapons production, 1942–1985. Unpublished doctoral dissertation. Carnegie Mellon University.

Taylor, Bryan C. 1990. Reminiscences of Los Alamos: Narrative, critical theory and the organizational subject. *Western Journal of Speech Communication* 54: 395–419.

——. 1993. *Fat Man and Little Boy*: Cinematic representation of interests in the nuclear weapons organization. *Critical Studies in Mass Communication* 10: 367–394.

——. 1997. Revisiting nuclear history: Narrative conflict at the Bradbury Science Museum. *Studies in Cultures: Organizations and Societies* 3: 119–145.

Taylor, Bryan C., and Brian Freer. 2002. Containing the nuclear past: The politics of history and heritage at the Hanford Plutonium Works. *Journal of Organizational Change Management* 15: 563–588.

Tuler, Seth. 2000. Forms of talk in policy dialogue: Distinguishing between adversarial and collaborative discourse. *Journal of Risk Research* 3: 1–17.

U.S. Department of Energy (USDOE). 1995, January. *Closing the circle on the splitting of the atom: The environmental legacy of nuclear weapons production in the*

United States and what the Department of Energy is doing about it. Washington, DC: DOE, Office of Environmental Management.
———. 2001, January. *Report to Congress on long-term stewardship.* Washington, D.C.: US DOE, Office of Environmental Management.
U.S. Department of Energy, Feed Materials Production Center (USDOE-FMPC). 1989, May 15. *RI/FS Community Meeting.* Fernald, Ohio. Appendix B of the Final Report FMPC–RI/FS Community Meeting of May 15, 1989 (1990, January 23).
———. 1989, October 24. *RI/FS Community Meeting.* Fernald, Ohio. Available from Spangler Reporting Services 513-381-3330.
———. 1990, May 22. *RI/FS Community Meeting.* Fernald, Ohio. Available from Spangler Reporting Services 513-381-3330.
———. 1990, June 12. *RI/FS-EIS scoping meeting.* Fernald, Ohio. Available from Spangler Reporting Services 513-381-3330.
———. 1990, August 16. *K-65 Silos Removal Action Community Workshop and Site-Specific Presentation.* Fernald, Ohio. Available from Spangler Reporting Services 513-381-3330.
U.S. Department of Energy, Fernald Environmental Management Project (USDOE-FEMP). 1993, June 22. *Community Meeting.* Fernald, Ohio. Available from the Public Environmental Information Center 513-648-3000.
———. 1993, October 21. *Community Meeting.* Fernald, Ohio. Available from Spangler Reporting Services 513-381-3330.
———. 1994, November 8. *US Department of Energy Public Meeting for Operable Unit 2 Proposed Plan.* Fernald, Ohio. Available from Spangler Reporting Services 513-381-3330.
———. 1995, January. *Community Relations Plan.* Fernald, Ohio.
———. 1995, August 8. *Community Meeting.* Fernald, Ohio. Available from Spangler Reporting Services 513-381-3330.
U.S. Department of Energy, Fernald Field Office. (USDOE-FN). 1993, November. *Public Involvement Program for the Fernald Environmental Management Project.* Fernald, Ohio.
U.S. Department of Energy, Fernald Area Office (USDOE-FN). 1999, June. *Responses to public comments on the environmental assessment for final land use at the Fernald Environmental Management Project.* Fernald, Ohio: U.S. Department of Energy.
U.S. Environmental Protection Agency (USEPA). 1990, May 9. *Public meeting regarding the U.S. Environmental Protection Agency—U.S. Department of Energy Consent Agreement for the Feed Materials Production Center site in Fernald, Ohio.* Fernald, Ohio. Available from Ace Reporting Services 513-241-3200.
Veiluva, Michael. 1995. Federal responsibilities and realities: An alternative view of the cleanup of the nuclear weapons complex. *Social Justice* 22: 126–139.
Waddell, Craig. 1996. Saving the Great Lakes: Public participation in environmental policy. Pp. 141–165 in *Green culture: Environmental rhetoric in comtemporary America*, ed. Carl G. Herndl and Stuart C. Brown. Madison: University of Wisconsin Press.
Weeks, Jennifer. 2000. Advice—and consent? The Department of Energy's Site-Specific Advisory Boards. Discussion paper 2000-14, Cambridge, Mass.: Belfer Cen-

ter for Science and International Affairs, ksgnotes1.harvard.edu/BCSIA/Library.nsf/pubs/advice&consent (accessed April 18, 2006)

Wells, James R., and Henry Spitz. 2003. Long-term stewardship of the environmental legacy at restored sites within the Department of Energy nuclear weapons complex. *Health Physics* 85: 578–584.

Wertsch, James V. 1987. Modes of discourse in the nuclear arms debate. *Current Research on Peace and Violence* 10: 102–112.

Williams, Walter L. 2002. *Determining our environments: The role of Department of Energy Citizen Advisory Boards.* Westport, Conn.: Greenwood Press.

Wills-Toker, Caitlin. 2004. Public participation or stakeholder frustration: An analysis of consensus-based participation in the Georgia Ports Authority's Stakeholder Evaluation Group. Pp. 175–199 in *Communication and public participation in environmental decision making*, eds. Stephen Depoe, John Delicath, and Marie-France A. Elsenbeer. Albany: SUNY Press.

3

Becoming Hanford Downwinders

Producing Community and Challenging Discursive Containment

William J. Kinsella and Jay Mullen[1]

Within days of the bombings of Hiroshima and Nagasaki in 1945, the U.S. government took steps to explain and contain the realities of nuclear energy. At the direction of General Leslie Groves, the military head of the Manhattan Project, Princeton University physicist H. D. Smyth had prepared a book-length report in advance of the bombings for public distribution immediately afterward. With presidential approval, the War Department released the report on August 12, three days after the destruction of Nagasaki (Rhodes 1995). Smyth opened the book's preface by declaring that "the ultimate responsibility for our nation's policy rests on its citizens" and that "they can discharge such responsibilities wisely only if they are informed" (Smyth 1945, vii). At the report's conclusion he stressed the need for open public discourse, acknowledging that scientists had ushered in a new age that posed new questions to humankind and asserting that "these are not technical questions; they are political and social questions and the answers given to them may affect all mankind for generations." In a Jeffersonian-Lincolnian tenor he emphasized that "in a free country like ours such questions should be debated by the people and decisions must be made by the people through their representatives" (226).

The Smyth Report, entitled *Atomic Energy for Military Purposes*, also contains a foreword by General Groves, whose remarks seem incongruent with those of Smyth. Groves warned sternly that "all pertinent scientific information which can be released to the public at this time without violating the needs of national security is contained in this volume," and that "no requests for additional information should be made to private persons or organizations associated directly or indirectly with the project." The gravity of this prohibition is clear in Groves' statement that "persons disclosing or securing additional information by any means whatsoever without authorization are subject to severe penalties under the Espionage Act" (v).

So the Atomic Age opened in paradox: Citizens must be informed participants in atomic matters, while national security limits their access to information. One effect of Grove's caveat was to obstruct the acquisition of information about what the report termed "unpredictable hazards" and "really troublesome materials" (121) associated with the industrial production of nuclear weapons. A particular hazard was airborne emissions of radioactive xenon and iodine, resulting from chemical reprocessing of irradiated reactor fuel to extract plutonium. The report acknowledged that "it must be established that the mixing of [these] radioactive gases with the atmosphere will not endanger the surrounding territory" (121), and recognized the importance of the "determination of biological effects" including "effects of radiation on living matter [and] metabolism" (125–126). A section entitled "radioactive poisons" observed that, if released, the fission products resulting from a single day's operation of a plutonium production reactor "might be sufficient to make a large area uninhabitable" (65). A reader might wonder how these concerns squared with claims that "the tolerance standards that were set and met were so rigid as to leave not the slightest probability of danger to the health of the community or operating personnel" (121). Thus the Smyth report established some of the foundational tensions structuring what Taylor, Kinsella, Depoe, and Metzler (2005) have called the "nuclear public sphere." What would be the public's discursive role in the face of national security's informational constraints and the elasticity of science, particularly with respect to issues of public health?

These paradoxes presaged the public controversy that is the context for our analysis, a controversy that emerged in the 1980s and continues to unfold six decades after Hiroshima and Nagasaki. As the nuclear age has evolved, officials have faced problems of increasing scope and complexity surrounding the containment of dangerous materials. In parallel, they have faced equally troubling problems surrounding their efforts to contain public discourse (Kinsella 2001; Taylor et al. 2005). Just as nuclear materials can be "unpredictable," and "really troublesome," so can be public discourse and the public itself. This chapter chronicles the emergence of a community linked by the principle of nuclear hazards, whose members have challenged and disrupted the containment of nuclear discourse. As self-identified "downwinders," the members of this community are not only linked by those hazards; in so naming themselves they have contributed to the discursive constitution of those hazards. Thus the downwinder community is simultaneously a product of nuclear materialities, a rhetorical agent in the constitution of those materialities, and a product of self-organization through rhetorical action.

The downwinder community comprises a number of more specific communities whose members identify with particular sites of nuclear production and testing (see Ball 1986; Caulfield 1989; Fradkin 1989; Gallagher 1993; Hacker 1998; Roff 1995; Taylor 1997; Udall 1994; Wasserman and Solomon 1982). These groups have forged a common identity; indeed, a slogan that circulates on websites and email messages proclaims that "we are all downwinders," stressing the ubiquity of nuclear fallout and expanding the range of identification.[2] We focus here, however, on the downwinders who claim connections to the U.S. Department of Energy's Hanford Reservation in Washington, the site of plutonium production for the Trinity atomic test, the bomb dropped on Nagasaki, and many of the approximately 70,000 nuclear warheads subsequently produced for the U.S. nuclear arsenal.[3] Plutonium production continued at Hanford through the end of the Cold War, raising troubling questions regarding toxic releases and regarding containment of public discourse. By examining the personal narratives, shared positionalities, and rhetorical strategies of the Hanford downwinders, we seek to illuminate a network of connections among materiality, discourse, identity, community, and democracy.

The Hanford downwinder community emerged autopoietically in response to a set of environmental threats, even as it acted discursively to construct and frame those threats. Following revelations in the mid-1980s about releases of radioactive materials into the environment surrounding Hanford, regional residents and former residents forged connections through their shared self-identifications as downwinders. Building on this shared identity, they developed rhetorical strategies that successfully connected them with downwinders from other regions, congressional policy-makers, and a broader public. In doing so the Hanford downwinders first challenged Cold War conventions of secrecy, and then challenged a technocratic framework that also served to contain public nuclear discourse. Grounding their claims in lived experience, local knowledge, and a humanistic ethos, the downwinders expanded the possibilities for democratic dialogue regarding technological choices and public policy.

Our analysis of this process acknowledges a range of published accounts of Hanford and its regional community (Dalton, Garb, Lovrich, Pierce, and Whiteley 1999; D'Antonio 1993; Gephart 2003; Gerber 1992; Goldberg 1998; Hales 1997; Hein 2003; Kaplan 1992, 2000; Kinsella 2001; Limerick 1998; Loeb 1982; Makhijani, Hu, and Yih 1995; Nussbaum, Hoover, Grossman, and Nussbaum 2004; Ratliff 1997; Ratliff and Salvador 1994; Sanger 1995; Taylor and Freer 2002). However, here we present a narrative that, while written in the third person, is grounded in the second author's experiences as a Hanford downwinder and a participant in the events we describe. Our narrative strives to be both analytical and anecdotal, grounding its validity in the actions, experiences, and names of actual people, thus mirroring the approach taken by the downwinders themselves. After establishing a communication-theoretical framework for our analysis, we present this narrative in sections corresponding to five moments in the development of the Hanford downwinder community: initial awareness, early self-organization, identification with a broader discursive and political community, engagement with the secretive and technocratic frames of public nuclear discourse, and the development of a public voice and policy influence. We then comment on the broader significance of the Hanford downwinders' experience, examining implications for the three problematics identified by Taylor et al. (2005): the nuclear public sphere; nuclear organizations, crisis, and change; and nuclear history, memory, and heritage. We believe that this case study has continuing relevance at a time of renewed public debate over the relationships among democracy, community, and government secrecy in a context of heightened national security concerns.

THEORETICAL FRAMEWORK: CONSTITUTIVE COMMUNICATION AND DEMOCRATIC DISCOURSE

Two analytical themes run through our account. First, we view the self-organization of the Hanford downwinders as a case study illuminating how an environmental risk and an associated, self-identified "risk community" are constituted mutually through communication. Second, we situate the downwinder community within the context of late and post–Cold War nuclear discourse, examining tensions among democratic openness, government secrecy, technical expertise, and local knowledge. These themes are closely related: We view the emergence of the downwinder community as a response not only to perceived environmental hazards, but also to constraints on democratic public discourse. We argue further that in both of these interrelated domains, the material and the discursive, the downwinders did not simply *respond* to prevailing conditions; they also actively *constituted* those conditions as social problems through communication.

Self-organization, Constitutive Communication, and the Production of Risk and a "Risk Community"

Kenneth Burke's concept of rhetorical identification (e.g., Burke 1969), as elaborated by Cheney (1983a; 1983b 1991), illuminates the self-organization of the downwinder community. Multiple, related forms of identification are evident in the downwinder narrative. Through communication with others, individuals who had not previously regarded themselves as downwinders began to do so, with important consequences for their values, practices, and personal and social identities. By adopting downwinder identities these individuals responded to two mysteries that pervaded their lives: mysterious personal and family health problems, and a mysterious government facility that had long dominated their regional landscape. An emerging downwinder narrative linked these previously unconnected phenomena together, engaging the powerful nuclear-discursive themes of mystery, potency, secrecy, and entelechy (Kinsella 2005). Consubstantiality and community emerged through that narrative, as individuals inferred that activities at Hanford had produced profound effects on their very bodies. As Burke (1969) suggests:

> A doctrine of *consubstantiality*, either explicit or implicit, may be necessary to any way of life. For substance, in the old philosophies, was an act; and a way of life is an *acting-together*; and in acting together [people] have common sensations, concepts, images, ideas, attitudes that make them *consubstantial* (21, emphasis in original).

In recovering the classical notion of substance as action, Burke characterizes consubstantiality as a shared *stance*, and indeed, the downwinders would come to share a stance and a way of life as they acted together toward the Department of Energy and other federal agencies. At the same time, however, a theme of material consubstantiality also grounds the downwinder narrative, manifested in their new attitudes toward the air they breathed, the food they ate, and the bodily effects they endured. Those shared, embodied materialities are both foundations for, and products of, the downwinders' emergent identifications.

Karl Weick's model of sensemaking also resonates with the downwinder narrative. Differentiating sensemaking from "other explanatory processes such as understanding, interpretation, and attribution," Weick (1995) proposes that sensemaking is "1) grounded in identity construction, 2) retrospective, 3) enactive of sensible environments, 4) social, 5) ongoing, 6) focused on and by extracted cues, [and] 7) driven by plausibility rather than accuracy" (17). Each of these properties is evident in the downwinder experience, and any of them could be developed in rich detail; here we can only sketch out the most basic connections. Consistent with Burke's principle of identification, we note the close connection between the downwinders' personal and group identities and their ongoing, social effort to make sense of their situation. This effort has been, and continues to be, a retrospective one, as downwinders make sense of previously inexplicable health problems and personal experiences. In doing so the downwinders have enacted a sensible environment of nuclear health hazards, which they believe to have prevailed during their childhood as they grew up—and grew ill—in the shadow of Hanford. By extracting cues on the basis of plausibility using local knowledge, embodied experience, and personal and community memory, the downwinders have posed a sensemaking challenge to the technocratic logic of the Department of Energy and other federal agencies. Although the accuracy of many of the downwinders' claims remains contested, the overall plausibility of their narrative has been a key factor—perhaps the primary factor—in its rhetorical force.

Weick's emphasis on plausibility rather than accuracy is consistent with the premises of narrative theory. Walter Fisher (1987) contrasts the "rational-world paradigm" that prevails in public policy discourse, which privileges technical expertise and field-specific standards of proof, with a "narrative paradigm" in which:

> The experts' stories are not at all beyond analysis by the layperson. The lay audience can test the stories for coherence and fidelity. The lay audience is not perceived as a group of observers, but as active, irrepressible participants in the meaning-formation of the stories that any and all storytellers tell in discourses about nuclear weapons or any other issue that impinges on how people are to be conceived and treated in their ordinary lives (72).

Elaborating the narrative theory of A. J. Greimas (1987), Cooren (2000) observes that "[n]arrativity can be thought of as the capacity for people to throw themselves into the future, to project something in order to fill a gap or restore an order that was initially disrupted" (60). Multiple disruptions of order are evident in the downwinder narrative: contaminants released from Hanford constitute pollution, defined by Douglas (1966) in an anthropological sense as "matter out of place"; this pollution has disrupted the health (the bodily order) and the lives (the social order) of Hanford's neighbors; and, importantly, these releases constitute a violation of the political compact between public officials and citizens.[4] This last disruption may be the most fundamental one within the downwinder narrative. In the view of many downwinders, although it is too late to repair past material releases or to prevent illnesses and deaths those releases may have produced, it is not too late to restore the democratic balance between government and citizens. Through that corrective action future transgressions can be prevented at Hanford and elsewhere, and future material harms can be prevented or ameliorated.

Thematically, the downwinder narrative is organized around this "basic tension" and the "quest" that it motivates; "subject and object are thus mutually defined" and "all the activities it implies *make sense*" (Cooren 2000, 60, emphasis in original). The narrative links sensemaking, stance, and action, enabling the production of a shared downwinder subjectivity and shared objects of concern. Those objects include material hazards, health disorders, and failures of democracy, all enacted on the basis of plausibility and extracted cues. Along with the shared downwinder subjectivity, a "mirror-subject" and a corresponding "mirror-narrative" (Cooren 2000, 62) emerge, identified with the Department of Energy and other federal agencies. These agencies have their own objects of concern, mirroring those of the downwinders, including the containment of discursive challenges and the preservation of discursive authority. In Burkean terms, a process of *association* links the downwinders, while a process of *dissociation* separates them from the federal government with which they had formerly identified (see Burke 1969; Cheney 1983a). This narrative and sensemaking analysis is consistent with the principles of symbolic convergence theory (Bormann, Cragan, and Shields 1996, 2001), in which group cohesion emerges from shared, social "consciousness-raising," shared "fantasy themes" (irrespective of their accuracy) and shared symbols including heroes, villains, enemies, and artifacts. However, these processes cannot be reduced to a symbolic dimension only; group cohesion and "narrative fidelity" (Fisher 1987) are also products of, and constitutive resources for, the downwinders' shared *material* experiences. Consistent with post-semiotic models of constitutive communication (e.g., Stewart 1995, 1996) the downwinder narrative problematizes simple distinctions between the symbolic and the material.

Taken together, the identification, sensemaking, narrative, and symbolic convergence perspectives support a constitutive view of communication as the organizing principle for an emergent downwinder community. Two perspectives from European sociology help us to address how material hazards and democratic failures, both key elements in the downwinder narrative, are also discursively constituted. In his analysis of "ecological communication," Niklas Luhmann (1989) observes that "society has . . . become alarmed as never before" (1) as "the system registers the effects of its own behavior upon the environment" (23). At such moments, communication constitutes environmental threats by selecting—or as Weick (1995) would say, enacting—those effects as objects of sensemaking and active response. Although the material status of those threats may be uncertain or contested—and often *because* of that state of affairs—they induce further communication and social action. Such is the case in the downwinders' confrontation with the nuclear establishment.

Similarly, Ulrich Beck (1992, 1994, 1995) argues that contemporary "risk society" is organized through two fundamental, sociological principles. Along with the familiar "scarcity" principle associated with the equitable distribution of limited resources, a situation of "reflexive modernity" entails the additional problem of equitably distributing the risks that inevitably accompany technological projects (see Flyvbjerg 2003). Within this framework individuals and groups occupy a range of "risk positions" related to their socioeconomic status, their political power, and their consequent, differential exposures to risks. Using Beck's terminology as a model, we view the Hanford downwinders as a self-identified "risk community" emerging autopoietically through identification, sense-making, and narrative action. Thus the downwinder narrative simultaneously constitutes material hazards, democratic tensions, and a community of affected members.

The Downwinder Narrative and the Nuclear Public Sphere

These links between risk, democracy, and community situate the downwinder narrative within the nuclear public sphere framework established by Taylor et al. (2005; see also Cox 2006, regarding the "environmental public sphere"). As Luhmann (1989) observes, contemporary public deliberation increasingly relies upon the languages and logics of specialized fields such as economics, politics, science, or law, rather than adopting any broader perspective; thus "system rationality increasingly loses its claim to be world-rationality" (Luhmann 1989, 138). Narratives such as that of the downwinders remain marginal within this system rationality (Fisher 1987), leaving public discourse "rhetorically attenuated" and "morally attenuated" (Pearce and Littlejohn 1997). These discursive boundaries often serve stakeholders poorly, as the interests and local knowledge of particular risk communities are marginalized or excluded (Fischer 2000; Flyvbjerg 1998, 2001; Peterson 1997; Richardson, Sherman, and Gismondi 1993; Williams and Matheny 1995; Wynne 1996). This tendency is perhaps most evident in the nuclear public sphere, a paradigmatic site of technocratic authority (Kinsella 2005).

Using Hanford as a case study, Kinsella (2001) draws an analogy between officials' efforts to contain nuclear materials and their efforts to contain public discourse, arguing that historically, discursive containment at Hanford has operated in two modes. Throughout most of the Cold War era, the principle of secrecy set forth so forcefully by General Groves provided an effective tool for containing public knowledge, and consequently, public discourse about Hanford. However, in the mid-1980s whistleblowers, investigative journalists, activists, independent scientists, and downwinders all challenged the principle of secrecy, producing a dramatic discursive rupture. Since that time a second form of discursive containment—present all along but now increasingly significant—has complemented the secrecy principle. This modality shifts the emphasis from *information* to *interpretation*, seeking to limit the range of voices that can speak with authority about the environmental and health risks posed by nuclear operations. Thus the downwinders' engagement with the Department of Energy and other federal agencies has pitted their local knowledge and plausible narrative against the technocratic narratives of epidemiology, quantitative modeling, and scientific authority. In that encounter, the downwinder narrative has challenged the monolithic authority of official, technocratic discourse while also adapting to that frame by incorporating forms of "popular epidemiology" (Brown 1987; Brown and Mikkelsen 1990; Clapp 2002) and "public expertise" (Kinsella 2002, 2004a). Thus the downwinders have made progress toward achieving the "trinity of voice" described by Senecah (2002), comprising access, standing, and influence in the democratic public sphere.

THE DOWNWINDER NARRATIVE: MATERIALITY, DISCOURSE, AND EMERGENT COMMUNITY

We now provide a representative version of the Hanford downwinder narrative, written by a self-identified member of that community. This section of the chapter forgoes analytical commentary, with the exception of our punctuation of the narrative into five moments: initial awareness, early self-organization, identification with a broader discursive and political community, engagement with the secretive and technocratic frames of public nuclear discourse, and the development of a public voice and policy influence. Although we view these moments as analytically distinct, we are not suggesting a "stage" model of organizational process; rather, the moments overlap chronologically and are ongoing themes as the downwinder community continues to evolve. Considered in the light of the analytical framework we have developed above, the downwinder narrative provides a case study of self-organization, constitutive communication, and rhetorical action in the nuclear public sphere.

From Containment to Questions: Material and Discursive Leakages

Friends and classmates from Couer d'Alene, Idaho, may have remembered Gertie Carder fondly when they received her class reunion announcements in the mid-1980s. Now known as Gertie Hanson, this homemaker discerned a disturbing pattern in their replies. Too many said they would not attend reunions because of poor health. She began analyzing responses from graduates of the years 1951–1954. Out of 242 respondents, 42 percent of the females reported thyroid problems that warranted medication. The incidence spiked to 52 percent for female respondents from the class of '53. Others reported cancers. To Hanson the numbers seemed inordinately high for people in their early fifties. Likewise, the newspaper's frequent obituaries of local classmates unsettled her.

Sixty-three had died of cancer by 1988, three of those from leukemia between 1948 and 1950. These anomalies haunted her in her kitchen and in her car until she heard about radioactive iodine releases from the atomic facility at Hanford, less than half as distant as Mount St. Helens, which in 1980 had deposited wind-borne ash in her yard. An activist group in nearby Spokane, Washington, had used the federal Freedom of Information Act to pry documents out of Hanford that alerted Hanson to the dangers of radioactive iodine.

Latent nuclear apprehension converged with a public context of growing environmental concerns to breach the decades of secrecy that General Groves and the Cold War had imposed upon atomic operations. Watergate, Love Canal, Agent Orange, falsified Cambodia bombing reports, Iran-Contra, The Pentagon Papers, and the Church committee had combined to undermine the unquestioning trust that some of the Depression generation had accorded their government. National security claims lost legitimacy with jaded Americans who raised environmental concerns only to be dismissed by officials who invoked secrecy. Health concerns in the 1960s had contributed to the enactment of an atmospheric test ban treaty. Protests such as those by Utah sheep ranchers, living and working downwind of the Nevada test site, had intensified and disseminated nuclear anxieties (Hacker 1998; Udall 1994). On the heels of Three Mile Island, a proposal to make Hanford a nuclear waste repository impelled the formation of the Hanford Education Action League (HEAL) by activists in Spokane concerned not just with a waste site, but with nuclear operations in general.[5] On the basis of disclosures by a whistleblower, a Seattle newspaper had reported an accidental radiation release at Hanford. In response a Unitarian clergyman and PhD chemist in Spokane, Will Houff, built upon public skepticism and apprehension to deliver a provocative sermon entitled "Silent Holocaust" in May 1984. One outcome was the formation of HEAL by September (D'Antonio 1993; Ratliff 1997; Ratliff and Salvador 1994). Abetted by information derived from worldwide antinuclear critics and researchers, and particularly by the efforts of a local chemistry professor, Allen Benson, the organization began to scrutinize eastern Washington's nuclear facility.

HEAL influenced the investigative journalism of Karen Dorn Steele of *The Spokane Spokesman Review*, which heightened public awareness of Hanford and intensified the pressure for more information about its secret operations. Assisted by a local farmer, Tom Bailie, Steele interviewed farm residents in the Hanford neighborhood about rumors of deformed livestock and discovered that like Gertie Hanson, Juanita Andrewjeski and Don Worsham had compiled their own morbid data on cancers and death. Augmented by Hanford documents acquired through Freedom of Information Act requests, Steele wrote about local health apprehensions. At a subsequent public forum at Spokane's Ridpath Hotel Hanford's manager, Michael Lawrence, responded to popular concern with assurances that the Department of Energy (DOE) would provide a summary of Hanford's past emissions. At a HEAL symposium that followed, Lawrence expanded his previous declaration and pledged to make primary documents available to the public. On February 27, 1986, the DOE released a 19,000-page trove of previously classified Hanford documents. HEAL researchers Tim Connor and Jim Thomas sifted meticulously through those documents, tracing a skein of releases and cover-ups which, when reported by Steele in the *Spokesman Review*, confirmed the fears of Andrewjeski and Worsham and amplified public concern. When the wire services relayed the information, Gertie Hanson connected the dots in the pattern of her own suspicions about her classmates' cancers.

The documentary record revealed not only accidental releases, but also deliberate ones. Releases in December 1944 and early 1945 could be attributed to wartime urgency and an absence of air filters, but the documents revealed that in 1949, a year after filters had been installed, they had been deliberately bypassed for experimental releases that spewed nearly 8,000 curies of radioactive iodine into the atmosphere over a two-day period (Gerber 1992).[6] Three Mile Island, by comparison, would later discharge 15 to 20 curies. The Smyth report's declared concern "to leave not the slightest probability of danger to the health of the community" (Smyth 1945, 121) had obviously been disregarded.

Steele's reporting of the "Green Run," as these experimental releases were dubbed,[7] intensified public awareness and concern. In turn, officials reacted. Oregon's and Washington's governors appointed a committee to assess the environmental significance of the releases. A few months later the federal Centers for Disease Control (CDC) addressed the need for health effects studies. The Colville tribe filed a lawsuit and the Department of Energy responded by initiating an emissions survey. The governors' committee recommended a similar "dose" study. The DOE merged the latter two studies into a single one, the Hanford Environmental Dose Reconstruction (HEDR), to be conducted by Pacific Northwest National Laboratories (PNNL), a government-owned, contractor-operated ("GOCO") facility adjacent to the Hanford site. A component of the DOE's national laboratory system, PNNL had been operated by the Battelle corporation since 1965. This arrangement meant that a DOE contractor would be investigating the effects of the DOE's Hanford operations, although a technical steering panel (TSP) of outside experts was appointed to monitor the project. Congress, meanwhile, acted with uncharacteristic alacrity and through the CDC funded yet another project, the Hanford Thyroid Disease Study (HTDS), to be conducted by the Fred Hutchinson Cancer Research Center in Seattle.[8]

Making Connections: Inferences, Identifications, and Self-organization

Inquiries from self-identified downwinders now inundated HEAL, which responded by sponsoring an informational conference in 1988. With more than 500 downwinders added to its mailing list, HEAL then acquired foundation support for a second conference in Spokane in May 1989. From Oregon, Washington, Montana, California, and Idaho they came. While downtown Spokane bustled with preparations for its Lilac Festival, over a hundred downwinders assembled in a campus lecture hall to be welcomed by HEAL's executive director, Lindy Cater, and to hear fellow downwinder Judith Jurji, the keynote speaker.

Fueled with fervor, this pale Seattle community college art teacher recounted her health history—fatigue, infertility, loss of skin pigment and sometimes even memory—which had caused her to wonder how, with so little strength and vigor, she could have descended from ancestors who crossed the plains. A student who grew up near Hanford had posed an explanation to her after one of their classes. Millie Smith had thyroid cancer and suggested Jurji see a physician to have her own gland checked. The resulting diagnosis was hypothyroidism. Additional checks confirmed the same condition in five members of Jurji's immediate family, although it did not occur among relatives who had not lived near Hanford. The heads of an inordinate number of listeners in the lecture hall nodded understandingly when she recounted this story. Many of them bore the surgical scar of thyroid removal, a downwinder identity badge that they, in grim humor, came to call "the Hanford necklace." Jurji's lament, "The rage I feel is unfathomable," resonated among many listeners who sat waiting for confirmation of their suffering from Dr. James Ruttenbur, the physician charged with initiating the CDC thyroid study. Some, however, harbored skepticism about Ruttenbur and his study.

Downwinders knew they had suffered. They knew thyroid disease, cancers and leukemia, stillbirths, fibromyalgia, birth defects, menstrual difficulties, and sterility. Of course these afflictions were not unique to downwinders. Some would have occurred had Hanford never existed. However, once apprised of the possible source of their misery, many had assumed that the government that had benevolently succored their families during the Great Depression, that had protected them from fascism and communism, that facilitated their home mortgages, secured their bank accounts, and guaranteed their retirements would recognize their suffering and fashion a protective agenda to spare others the horrors they had endured. Their suffering at the hands of their beneficent government must have been a grievous mistake, a colossal error. Yet their misery could serve a noble purpose: Knowledge derived from their experience could provide insights and information that could help protect future generations.

When they heard about the government's intention to study radiation effects they had offered themselves as prime examples, inadvertent guinea pigs, as it were, whose medical records would repay their humane government for its Depression-era generosity. So amid visions of stethoscopes, x-rays, and blood tests they had penned poignant letters volunteering as subjects for the CDC's studies. In turn they had received impersonal, almost dismissive form letters, closed with reproductions of James Ruttenbur's signature. The message was, in effect, "don't call us; we'll call you." Some seethed when he strode down the aisle to address the downwinder gathering.

Ruttenbur, of course, had no notion of their resentment. Trained as a physician, steeped in the health sciences paradigm, he endeavored to explain the rudiments of epidemiology. Data from a pilot study would determine if a larger study was warranted, he said. Hearing that, some listeners groaned, convinced that their health horrors alone provided sufficient warrant. Ruttenbur averred, rationally, that surveying pre-selected people would invalidate any epidemiology. He intended to weigh probabilities before postulating Hanford's effect on regional health. Downwinders, meanwhile, assumed that since the government had acknowledged that radiation affected health, effects were what should be studied. Their scars and medical bills evidenced Hanford's effects. Ruttenbur seemed to be discounting the obvious. His measured, logical, factual presentation intensified their resentment about the impersonal form letter with his reproduced signature. He, and by extension, the government, apparently had no sensitivity to their suffering.[9]

The contrast between Ruttenbur's plodding logic and the downwinders' emotionalism anticipated the two rhetorical strategies that would clash as the downwinders engaged the nuclear establishment. Grief, confusion, perplexity, anger, sympathy, resentment, tears, resignation, and resolve rang through the afternoon group and plenary discussions. At adjournment a coterie reconvened over drinks, then dinner, determined to formalize an organization. Newly acquainted individuals found it difficult to focus on organizing and mission definition, but clarity and direction began to emerge as June Stark Casey described her appearance, three months earlier, with Fernald, Ohio downwinders before a congressional committee in Washington.

Members of Congress had listened as Casey spoke of her total hair loss (like that of 127 Ukrainian children in a town 230 miles downwind from Chernobyl), her thyroid deficiency and nodules, her skin cancer, breast surgery, and her miscarriage in 1970. For a month she had carried a fetus she knew to be dead because her physician feared that induced labor would kill her. "You have no hair?" asked another downwinder. A tear rolled down her cheek as she fixed him with a grateful smile for not recognizing that she wore a wig. Then another acknowledged that she too had been totally bald during part of her primary school years, making her a playground freak, although her hair eventually returned. The evening was replete with such emotional disclosures.

HEAL's researcher, Jim Thomas, and Jurji's student, Millie Smith, also had appeared with Casey before the congressional committee. In a soft voice, all the softer because a vocal chord was paralyzed after her thyroid removal, Smith had told the nation's legislators, "Due to my health problems I've lived in poverty most of my life . . . There have been times when all I had to eat was rice and popcorn for months . . . I was homeless . . . no one would rent to me as I didn't have a job." She described feeding her daughter by relying on upon food pantries and the Salvation Army and spoke of Andrewjeski's "death mile" where, she said, "Twenty-four of twenty-eight families have been afflicted with cancer, thyroid problems, and serious birth defects such as no eyes, no skull, no hips." She then produced a letter from the Hanford Operations Office that stated:

> We have found nothing to indicate that your illness was caused by the negligent or wrongful act or omission of any employee of the United States while acting within the scope of his or her office or employment as required before a FTCA claim can be allowed. Your claim therefore is denied.

Such anguished anecdotes circulated around an oval dinner table as virtual strangers, bound together only by their Hanford connection, forged an organization. Their mission was not defined easily. Some wanted to campaign for compensation while others believed that the issue of personal money would dilute appeals for medical monitoring. Some believed that their foremost goal should be witnessing, stressing their personal miseries in hopes of forestalling future tragedies. Others simply wanted the truth about the whole sordid matter. Still others wanted an apology from the government and from those responsible for their misery and for withholding vital information. Consensus came slowly. Finally they affirmed Millie Smith's entreaty before Congress three months earlier when she had pleaded for "justice for all downwinders." Smith had called for the right to sue (to be achieved by repealing the Warner amendment that precluded nuclear facility lawsuits), for medical care and monitoring "as they have in Japan," and for "a toll free number where downwinders can get accurate information." We want the "full truth," she said. Her emotional congressional appearance personified the ethos of the Hanford Downwinder Coalition (HDC) as it organized itself. Her assertion that "[t]wo years ago I was fighting for my life, today I'm fighting for the life of all of us and for the life of our land," emerged as the community's credo.

Expanding Identifications

In the court of public opinion downwinders now sought to convince the jury, the citizenry, that Hanford operations not only had victimized them but also posed a continuing threat to everyone's health. Hanford's defenders, in turn, sought to dismiss downwinder concerns as hysterical fear-mongering, scientifically and technically unsound, pandering to the public's latent, irrational nuclear apprehensions. Hanford's errors had been acknowledged, but no conclusive evidence demonstrated that its operations had adversely affected anyone's health, said the facility's supporters. Excessive radiation *could* have, perhaps, but there was no proof that it *did*. Hanford's advocates invoked science; downwinders invoked emotion and lived experience. Neither alluded to the ultimate nuclear option, litigation, a risky and unpredictable path that remained the unspoken alternative to civic engagement.

Although financially deficient, downwinders were not without intangible resources. They already enjoyed a measure of support within the government. Three months before they organized, Ohio Congressman Thomas Luken had alleged that the DOE "had systematically ignored, underfunded, and lied about its controls or lack of controls" at Ohio's Fernald site.[10] In June 1989 the FBI, at the behest of the EPA, seized the DOE's Rocky Flats nuclear facility in Colorado to stop operations that belied management's health and safety assurances there (Lodwick 1993). Before summer's end Congress also held hearings on the Radiation Victims' Fair Treatment Act, responding to disturbing evidence of medical experiments conducted on patients, prisoners, and other vulnerable subjects, in many cases without informed consent. The hearings inspired Massachusetts' Ed Markey to forward to Energy Secretary John Herrington a staff report entitled *American Nuclear Guinea Pigs: Three Decades of Radiation Experiments on U.S. Citizens* (U.S. House of Representatives 1986). Downwinders were encouraged that Markey urged the Secretary "to make every practicable effort to identify the persons who served as experimental subjects, to examine the long-term histories of subjects for an increased incidence of radiation-associated diseases, and to compensate these unfortunate victims for suspected damages." Although the experimental subjects were not necessarily downwinders, these revelations raised the level of public concern regarding radiation hazards and the government's nuclear decision making.

Confronting Technocratic Conventions

Congressional responses and the events at Rocky Flats heartened downwinders in 1989, but they realized that unlike atomic workers and veterans, they possessed no formal status beyond their incorporation. They were self-identified victims without official recognition, whose claims to legitimacy hinged largely upon the results of the dose survey and the thyroid study. If either concluded that radiation exposure from Hanford was indeterminate or insufficient to cause widespread health effects, then their downwinder status would be discredited, if not destroyed. Not surprisingly they focused passionately on both projects.

The DOE had conceived the dose reconstruction project in controversy when it commissioned Battelle Laboratories to conduct it, for Battelle had accumulated a long history of public skepticism. In 1964 the DOE's predecessor, the Atomic Energy Commission (AEC), had commissioned Thomas Mancuso of the University of Pittsburgh to survey cancers at Hanford, presuming that he would affirm atomic worker safety. When his preliminary findings indicated otherwise, his contract was canceled. He and his colleagues eventually reported a death risk far higher than prevailing assumptions (Mancuso, Stewart, and Kneale 1977). Meanwhile, the official who initially recommended his termination left the AEC to work for Battelle and was awarded Mancuso's lapsed contract. In a 1978 letter to Secretary of Energy James Schlesinger, congressional representatives Paul Rogers and Tim Lee Carter concluded that the reasons for firing Mancuso were "not supported" and that transferring his research to Battelle was of "highly questionable legality" (Wasserman and Solomon 1982, 122). From the Mancuso affair downwinders concluded that the logic of science is disputable, not immutable—a useful argument when appealing to the public—and that Battelle warranted close scrutiny.

Downwinders themselves did not typically presume to contest the establishment's scientists, but in public forums they enthusiastically broadcast the conclusions of other scientists who did. A key opportunity to do so came when Physicians for Social Responsibility published *Dead Reckoning: A Critical Review of the Department of Energy's Epidemiological Research* (Geiger and Rush 1992), asserting that "there is evidence extending over many decades of intermittent administrative attempts . . . to suppress evidence suggesting health risks, to intimidate some epidemiologic and environmental investigators, and to highlight reassuring findings while downplaying or denying risks," and concluding that the "epidemiology program is seriously flawed" (13).[11] Citing such sources enhanced the downwinders' legitimacy as both knowledgeable critics and nuclear victims.

Increasingly, the dose reconstruction project emerged as a lightning rod that drew the downwinders' coalesced energies. John Till, a retired naval officer, chaired the Technical Steering Panel charged with oversight of the project. TSP members were selected for professional acumen, not interpersonal skills, and Till later acknowledged that the course of their project taught them much about public relations.[12] At an October 1989 meeting at Jantzen Beach, on the Columbia river in Portland, Oregon, the tension was palpable. Till opened the meeting with a welcome to the public and an assurance that the panel valued its participation, yet individual members bristled at participants whose questions impugned the project's integrity. A substantial part of the analysis depended upon access to Hanford documents, but Hanford officials could invoke national security to withhold records from public scrutiny. A simple question about the physical location and public access to documents affronted one panelist, who assumed that it implied a cover-up. And although they reflected genuine public concern, questions about topics such as Nevada weapons tests, sick uranium miners, or Hiroshima exposures likewise exasperated the panel, for they had no direct relevance to the design of the HEDR project. References to dosimetry, ions, isotopes, rads, and picocuries confused some listeners and convinced others that the panel was invoking scientific obscurantism to discredit those who presumed to challenge it. When a frail, grey-haired woman began posing a question about her atomic worker husband's death, Till brusquely dismissed her by declaring that such matters were beyond the project's purview. The tear on the widow's cheek confirmed to the downwinders that they should not look to HEDR personnel and their project for humane support.

Finding a Voice: Public Legitimacy and Political Influence

Thus the downwinder community, initially inchoate and directionless, developed a strategy of engagement as human, and humane, participants in the nuclear dialogue. Although excluded from the dose reconstruction project, downwinders infiltrated various other public advisory organizations for government programs. Participation in working groups and advisory boards associated with the Hanford Health Effects Subcommittee (HHES), the Hanford Health Information Network (HHIN), the Hanford Health Information Archives (HHIA), and the Agency for Toxic Substances and Disease Registry (ATSDR) provided them not only with platforms, but with opportunities to travel at government expense over the vast distances that separated them to attend meetings where they could share mutual support and strategize. Usually they left the task of contesting scientific assertions to their technical allies, while seeking to use their own ethos as victims to generate sympathy, raise public awareness, and chip away at the legitimacy of the nuclear community. Their presence would inform or remind the public of the Green Run's implications, supplemented by reminders of Three Mile Island, Chernobyl, and "nuclear guinea pig studies." They proved to be vivid anecdotalists not only at public meetings but for news media ravenous for reportorial color. Downwinder Lois Camp, for example, uncovered an early Hanford document (Iodine-131 deposition 1946) that made a mockery of the scientific procedures for radiation-monitoring. In a virtual Keystone Cops routine, troops had bounded about pastures north of the reservation in jeeps, cornering recalcitrant cows so they could assess thyroid exposures with Geiger counters pressed to the animals' throats. "This is the vaunted scientific method?" downwinders would ask derisively.

Mesa, Washington, farmer Tom Bailie, dubbed "Mr. Downwinder" by Judith Jurji's husband, proved to be the downwinder raconteur par excellence. After helping Karen Dorn Steele gather information for her *Spokesman Review* articles, he penned his own op-ed for the *New York Times*, then appeared to speak wherever he could find an audience, from radio programs to the National Association of Radiation Survivors meeting in San Antonio, Texas. In a 1990 NOVA television program entitled "The Bomb's Lethal Legacy," he placed his booted foot on a mechanical harvester, tipped back the brim of his cowboy hat, and with Will Rogers folksiness described "the death mile" for a nationwide audience. In the cab of his pickup truck, he recounted for NOVA's camera the catalogue of illness in 21 of 23 families while he drove along that mile across the river from Hanford.

Meanwhile, other downwinders augmented pathos with logos in the form of empirical insights, methodological critiques, and local knowledge. Ida Hawkins, a homemaker and downwinder from Ellensburg, Washington, sifted through the turgid pages of the HTDS protocol and saw that CDC epidemiologists proposed comparing the irradiated neighborhood around Hanford with the purportedly unaffected community of Ellensburg, fifty miles to the northwest. Although prevailing winds usually blew from west to east, diffusing fallout in that direction, Hawkins knew that they sometimes shifted, so that Ellensburg too was exposed. Conclusions about exposure, then, would compare two exposed areas, not one exposed and an unexposed control. Alerted by Hawkins, available downwinders in Oregon converged upon a thyroid study public meeting in the basement of Portland's First Presbyterian Church, joining other watchdog groups including Columbia River United, which was concerned about radioactivity in the river. At the meeting's outset the epidemiologists declared that they intended to answer any and all questions, even if it took all night. After downwinders registered their concerns over the selection of Ellensburg and while others were expressing their anxieties about the contamination of the Columbia, which was not within the purview of the study, panel members began sneaking glances at their wristwatches. When pressed, they admitted that they had to catch a scheduled plane flight. Attendees groaned at the realization that the epidemiologists had never intended, as they had declared, to stay all night if necessary. A reminder of their stated commitment annoyed them while adding to public doubts about government sincerity, and by extension the integrity of their commitment to the public, even though they had acknowledged the validity of Ida Hawkins's concern.

As the downwinders' visibility and strategic sophistication grew, so did their influence in the national capitol. Senator Mark Hatfield of Oregon and Representative Thomas Foley of Washington shared the concerns of their eastern colleagues Luken and Markey about nuclear operations and the possible consequences for constituents. Thus section 3138 of the National Defense Authorization Act for fiscal year 1991 (Public Law 101-510) appropriated three million dollars for "programs for persons who may have been exposed to radiation released from the Hanford nuclear reservation," joining sections 3139-3141 which called for "payments for injuries believed to arise out of atomic weapons testing," patterned after the Radiation-Exposed Veterans Compensation Act. The state health departments of Washington, Oregon, and Idaho, which monitored the resulting programs, declared to Congress that "honesty, openness to differing points of view, and willingness to answer questions" would be among their "operating principles," that "accurate information will be produced that acknowledges emotional as well as physical health effects," and that "persons who may have been exposed to radiation released from Hanford" would be involved "in a voting capacity on all project advisory boards." State health officials appointed downwinder activists Gertie Hanson of Idaho, Judith Jurji of Washington, and Carol Baker of Oregon to their boards. Within two years, downwinders had evolved from unorganized inchoates to congressionally acknowledged stakeholders.

Always skeptical of both the DOE and the dose reconstruction project, at their organization's earliest stages downwinders had also weighed the prospects of litigation. So did Seattle attorney Tom Foulds, a World War II combat veteran who suspected his deceased wife had been a Hanford victim. He, too, doubted the dose project's integrity. After a series of negotiations with Judith Jurji acting as the primary downwinder spokesperson, Foulds and the downwinders agreed to collaborate on their own parallel study, the R-11 survey, sponsored by downwinders but largely funded by Foulds. Representing over 5000 litigants, in 1990 Foulds and a consortium of attorneys filed two lawsuits, *In re. Berg* and *Lumpkin et al. v DuPont et al.*, against the Hanford contractors who did not enjoy Warner amendment immunity from litigation (although the federal government pays those contractors' legal expenses). In 1998 downwinder and California attorney Tricia Pritikin filed another suit, seeking to compel federal funding for a medical monitoring program. Pritikin, a member of the Hanford Health Effects Subcommittee advisory board and chair of the Hanford Health Information Archives advisory board, is the daughter of a Hanford engineer who died of cancer. Like Jurji, she believes that radiation exposure has damaged her thyroid and caused her the loss of skin pigmentation.

In 2002 the U.S. Supreme Court disallowed Pritikin's case. However, despite a judge's delay to await the publication of the thyroid study conclusions and the removal of numerous litigants, followed by multiple motions and an initial dismissal in 1998, in 2002 the Ninth Court of Appeals reinstated a consolidation of the *Berg* and *Lumpkin* cases (*In re Hanford nuclear reservation litigation*).[13] In April 2005, sixteen years and eleven months after the downwinders first convened in Spokane, a federal jury began weighing reams of studies and hours of expert testimony, as well as the testimony of six "bellwethers," three of them with cancers, chosen from 2,300 downwinder plaintiffs. In pretrial discovery, previously classified documents indicated that the DOE and the Justice Department had set up the $27 million dose reconstruction project not to ascertain truth but to provide for "litigation defense." That revelation, by extension, cast doubt upon the thyroid study which utilized the dose reconstruction as a basis. The downwinders' persistent skepticism of both studies was validated. After three weeks the jury awarded over $500,000 to two of the three cancer victims, while concluding that no causal link had been demonstrated for the third cancer sufferer and the three bellweather litigants with autoimmune diseases. Despite these mixed findings, many downwinders claimed a victory. A citizen jury had affirmed what officials had spent almost two decades denying: Hanford operations had damaged public health.

In her 1989 appeal to the congressional Hazardous Materials Subcommittee, which became the basis for the Hanford Downwinder Coalition's activist agenda, Millie Smith had called for more public knowledge about Hanford's operations and effects. In putting that issue on the national agenda, and in their subsequent influence on the programs that resulted, downwinders have helped impart such knowledge. She had also pled for more "accurate information" and a toll-free number for public inquiries. Downwinder activity assisted in realizing these goals. Smith had also asked for an apology from the government, and while that has not been forthcoming, President Clinton's apology for the Tuskeegee syphilis experiments suggests that that it may be attainable. Her call for health care, compensation, and a repeal of the Warner amendment to enable litigation are, as yet, unrealized, but downwinders remain active on those issues. Thus a small and determined group, the Hanford Downwinder Coalition, has engaged and transformed public discourse. Their activism has infused color and intensity into the public nuclear dialogue, has challenged and expanded its rhetorical conventions, and has influenced policy. Meanwhile, Hanford downwinders Gertie Hanson and Carol Baker, who contributed so much of their time and energy to these transformations, have died of cancer. Their voices live on as the dialogue continues.

ANALYSIS: SELF-ORGANIZATION, DEMOCRACY, AND CONSTITUTIVE COMMUNICATION

We now provide a commentary on the Hanford downwinder narrative, grounded in the analytical framework we have outlined above. We organize our discussion according to the three problematics identified by Taylor et al. (2005): the nuclear public sphere; nuclear organizations, crisis, and change; and nuclear history, memory, and heritage. Taylor et al. consider "the discursive legacies of Cold War nuclear weapons production to be a site of social struggle: They are simultaneously invoked, contested, and reconstructed in communication among and between stakeholders" (379). Consistent with that view, we regard the Hanford downwinders as a self-organizing stakeholder community that has contributed significantly to public discourse regarding the past, present, and future of nuclear weapons production. Our analysis examines how those contributions have constituted and transformed linkages among material hazards, public discourse, individual and group identity, community, and democracy.

The Downwinder Community and the Nuclear Public Sphere

Accounts of communication surrounding the U.S. nuclear weapons production complex consistently observe that a radical challenge to a prevailing culture of secrecy emerged in the 1980s. Geiger and Rush (1992) provide one such account:

> The wall of secrecy did not begin to crumble until 1986. . . . Driven by public concern in the wake of the Chernobyl reactor accident, then Secretary of Energy John Herrington asked the National Academy of Sciences to assess the safety of the DOE reactors that produced nuclear weapons materials . . . . [The resulting] report triggered a cascade of further revelations by other government agencies, Congressional oversight committees, and investigative journalists, with far-reaching consequences (22).

Geiger and Rush identify the April 1986 Chernobyl accident and the subsequent reactor safety inquiry as the pivotal moment in the erosion of secrecy. Silverman (2000) points, instead, to a controversy over environmental releases of uranium at the Fernald site in Ohio, which began with a public announcement by DOE officials in November 1984, as the "first domino" (265) in a chain of challenges to the production complex (see Hamilton, in this volume). Both of these events, however, were preceded by a lawsuit filed in 1983 in connection with the DOE's operations at the Oak Ridge facility in Tennessee, which led to an April 1984 ruling holding the Department accountable under two environmental laws (see Gephart 2003; Taylor et al. 2005). Reverend Houff's sermon in May 1984, the formation of HEAL four months later, and the release of Hanford documents in February 1986 all followed the Oak Ridge lawsuit and preceded Chernobyl and the U.S. reactor inquiry. Thus something larger than any single event, and extending beyond any particular DOE facility, was under way. Together, in the broader contexts of a devolving Cold War, public environmental and nuclear anxieties, and a declining level of trust in government, events at Oak Ridge, Hanford, Fernald, and Rocky Flats constituted a discursive rupture no less consequential than the material releases that plagued the Department of Energy.

This notion of rupture does not imply that secrecy went unchallenged prior to the 1980s, or that it was a monolithic principle through which nuclear discourse had been contained (see Kinsella 2001, 2005). As early as 1948 an Atomic Energy Commission report (Williams 1948) had provided a "nationwide critique of the nuclear weapons complex" and "identified serious concerns about waste management at all sites" (Gephart 2003, 2.8). A year later the Columbia River Advisory Group, including officials from the U.S. Public Health Service and the Oregon and Washington state governments, was established to address the impact of Hanford's operations on the regional environment. The group's first press release "set off a flurry of controversy" regarding the implications of informing the public about issues viewed as both security-sensitive and technically arcane (Gephart 2003, 2.19). In 1973 a storage tank at Hanford leaked 115,000 gallons of high-level liquid waste into the ground, drawing national attention (Gephart 2003; Walker, Gould, and Woodhouse 1983).[14] Nevertheless, as Silverman (2000) notes, a "closed, expert-driven model enabled Hanford's managers to develop and maintain a consensus of environmental safety even as increasing evidence indicated weaknesses in the site's waste management practices" (quoted in Gephart 2003, 2.19). Thus beginning in the early years of nuclear weapons production, and throughout the decades that followed, the principles of secrecy and expertise interacted as complementary resources for containing discourse. Those principles, in turn, were warranted in a broader discursive formation that linked science, technology, government authority, and national security in powerful but historically contingent ways (Kinsella 2005).

Initially, the events that took place across the weapons production complex during the 1980s foregrounded issues of secrecy. As exemplified by the efforts of HEAL, journalist Karen Dorn Steele, and the Hanford downwinders, concerned groups and individuals struggled for basic access to information regarding nuclear operations, using techniques ranging from Freedom of Information Act (FOIA) requests to aggressive questioning of officials at public meetings (D'Antonio 1993). With the discursive rupture of the 1980s, however, a new set of challenges emerged as stakeholders found themselves confronted with enormous quantities of highly technical information. Struggles to define the meaning of information became as salient as struggles over access to information. Information management, technical competence, and interpretive authority became key issues of concern (Kinsella 2001).[15]

These developments led to the emergence of multiple forms of "public expertise" (Kinsella 2002, 2004a) within the Hanford stakeholder community; here we identify three such forms. First, HEAL mobilized its own technical competencies through the work of Reverend Houff, Alan Benson, Tim Connor, Jim Thomas and others. As Gephart (2003) notes, although HEAL has now disbanded some former members continue to play important roles in the ongoing dialogue between the public and the Department of Energy. Second, downwinders including Juanita Andrewjeski, Tom Bailie, Gertie Hanson, Ida Hawkins, and Don Worsham deployed local knowledge about family health histories, apparent clusters of illnesses suffered by people and animals, food consumption practices, local wind patterns, and illnesses among former residents who had moved away from the Hanford area. In compiling their own morbidity and mortality data and providing contextual information useful in their interpretation, the downwinders enacted the kind of "popular epidemiology" that Brown and Mikkelsen (1990; see also Brown 1987; Clapp 2002) have examined at other settings where releases of toxic chemicals were linked to issues of public health. These two forms of public expertise—stakeholders' own technical competencies and local knowledge unavailable to most official experts—were instrumental in the downwinders' challenges to the dose reconstruction project and the thyroid study.

A third form of public expertise consists of technical assistance and scientific authority provided to stakeholder groups by external, nongovernmental organizations sympathetic to their interests. The Institute for Energy and Environmental Research, a nonprofit organization committed to providing local communities with technical support, has been active at Fernald, Hanford, and other Department of Energy sites (Gephart 2003; Kinsella 2002). Physicians for Social Responsibility, the source of the influential *Dead Reckoning* report, has continued its critique of official epidemiology and its advocacy for stakeholder health concerns. In a collaboration linking downwinders with physicians, an epidemiologist, a nuclear scientist, and a database specialist, the Northwest Radiation Health Alliance has challenged the conclusions of the Hanford thyroid study using data from a community-based participatory health survey (Nussbaum et al. 2004). Alliance members argue that the participation of downwinders in the survey's design and distribution has enhanced its value as a tool for identifying anomalous health effects, while also increasing downwinders' competencies for self-advocacy, thus providing both scientific and social benefits. That argument extends methodological innovations emerging within the participatory research movement in the social sciences (e.g., Fischer 2000; Flyvbjerg 2001; Laird 1993) to the environmental and health sciences domains.

Public expertise, in any of the three forms described above, operates primarily within the scope of Luhmann's system rationality or Fisher's rational-world paradigm, vying for authority with alternate technocratic narratives offered by actors such as the Department of Energy. While using their own and their allies' public expertise to compete within that framework, however, the downwinders also posed a more radical challenge to the framework itself. Like other affected parties in a range of environmental and health controversies, the downwinders emphasized plausibility, coherence, narrative fidelity (Fisher 1987) and autobiographical rhetoric (Hope 2004) as alternatives to technical logic. This two-pronged rhetorical strategy was effective not only in engaging a broad range of audiences, but also in foregrounding the mutual relevance of technical and non-technical narratives. In this regard we situate the efforts of the Hanford downwinders within a broader genre of stakeholder challenges to the technocratic norms that prevail in public discourse surrounding environmental, energy, and sustainable development issues (cf. Fischer 2000; Peterson 1997; Richardson, Sherman, and Gismondi 1993; Taylor 1997; Wynne 1996). We wish to emphasize, however, the constitutive dimensions of such challenges, which do more than expand the boundaries of discourse surrounding preexisting issues. Until the downwinders and their allies raised questions about the potential health effects of Hanford's emissions, those effects were rarely a topic for public dialogue. Together, secrecy and technocratic authority constrained not only *how* such topics were discussed, but more fundamentally, *whether* they were discussed at all. In that sense, the downwinders' communicative action constituted a set of health risks and a corresponding set of democratic failures as public, social problems. As the downwinders connected their own physical symptoms to an emerging narrative of organizational errors and irresponsibility, they themselves became a visible symptom of those problems within the body politic.

Nuclear Organizations, Crisis, and Change

In constituting those social problems, the downwinders also helped constitute the late– and post–Cold War organizational crisis experienced by the Department of Energy and its nuclear weapons production complex. That crisis was, and continues to be, a material crisis, a managerial crisis, and a legitimation crisis. Although the term "crisis" suggests a sudden onset and a bounded duration, this one has been a half-century in the making and its legacies may persist for decades (the planned duration of the DOE's environmental cleanup program), centuries (the time frame for significant environmental and human health risks to be managed through a system of "long-term stewardship" and "institutional controls"; see Taylor, in this volume), and even millennia (the half-lives of some of the dangerous radionuclides). Kaplan (1992) has characterized the situation at Hanford as a "slow motion emergency," and that phrase applies as well to the situation throughout the entire nuclear weapons production complex.

Responding to the discursive rupture that swept through the complex in the mid-1980s, following the transition to a new presidential administration in 1993 the DOE embarked upon a self-described "openness initiative" under the leadership of Energy Secretary Hazel O'Leary (see Kinsella 2001; Taylor et al. 2005; USDOE 1995). This effort can be interpreted variously as an image-restoration strategy, an attempt to regain discursive control through reframing, a genuine commitment to open public discourse, or some combination of these elements. Regardless of the underlying intentions, the openness strategy had important effects, some of them irrevocable. Releases of huge quantities of information, a range of formal commitments to informational programs and dialogue with stakeholders, and the creation beginning in 1994 of site-specific advisory boards as forums for stakeholder involvement changed both the content and the context for the agency's interactions with the public. Although many stakeholders believe that the 1990s commitment to openness has now been reversed significantly, its legacies include greater public awareness of the environmental problems surrounding the weapons complex, continuing stakeholder expectations for information and involvement, and a more diverse and sophisticated array of active stakeholders.

In challenging the prevailing technocratic framework for deliberation, the downwinders and other stakeholder groups have also linked the DOE's organizational crisis to a more general crisis facing scientific and technological institutions, particularly those engaged in nuclear operations (see Kinsella 2004b). Potentially, a fourth category of public expertise resides within government or government-affiliated institutions charged with the mission of conducting science in the public interest. In this regard, however, the public advocacy roles of institutions can be ambiguous and contested, as in the case of Battelle Laboratories' management of the Hanford dose reconstruction project. After downwinders challenged Battelle's close relationship with the DOE and their concerns were validated by a congressional committee, funding for the project was transferred to the Centers for Disease Control, a federal institution more geographically and structurally distant from Hanford. Even then, however, Battelle continued as the research contractor. Downwinders remained skeptical of the project's objectivity due to Battelle's history of close association with the DOE, its geographic and social connections to Hanford, and its aspirations for continuing DOE contracts. Thus rather than enhancing the legitimacy of the dose reconstruction project, the funding transfer contaminated downwinders' perceptions of the CDC and its thyroid disease study, and their skepticism was further exacerbated by the thyroid study's methodological reliance on results from the dose reconstruction. Following the thyroid study's controversial preliminary report, released in January 1999, downwinders fought successfully for a review by the National Academy of Science (NAS). That review (National Academy of Science 2000) validated a number of their concerns (see Kinsella 2001), and most downwinders remained dissatisfied with the inconclusive findings of the eventual, final thyroid study report (Davis, Kopecky, and Hamilton 2002).

This episode features five federal actors—PNNL (operated by Battelle), the DOE, the congressional committee, the CDC, and the NAS—engaged in a complex negotiation for technical and administrative authority. Although all of these institutions are public in principle, downwinders questioned the degree to which at least three of them represented public interests. Here, as in the case of the post-Chernobyl reactor safety review, the NAS was widely regarded as a higher-order technical authority and a more trustworthy public advocate, although some stakeholders and public-interest organizations have challenged that perception (e.g., Center for Science in the Public Interest 2006). Other struggles of this type are manifested in the EPA's 1989 challenge to the DOE's Rocky Flats facility and in the complex, ongoing regulatory arrangements at the DOE's various nuclear cleanup sites. At Hanford, for example, both the EPA and the state of Washington maintain technical staffs in support of their regulatory authority under the conditions of a consent decree known informally as the "Tri-Party Agreement" (Hanford Federal Facility Agreement 2003). A number of neighboring Tribes and the nearby state of Oregon also have their own technical staffs devoted to Hanford issues. With the possible exception of the Tribes, who assert a government-to-government relationship in their interactions with the DOE, all of these actors might be categorized as loci of public expertise. However, disputes such as the one with the downwinders often complicate definitions of the public, the public interest, and public expertise. The dual principles of secrecy and contested expert authority are recurrent themes in such disputes.

An addition theme of organizational crisis and change, of course, surrounds the downwinder community itself. Crisis and change are evident in the downwinder narrative in connection with resource constraints, geographic dispersal of members, internal disagreements over interpretations of the situation and over strategic choices, and ongoing adaptation to the problem of framing a plausible public challenge to the dual authority of government and science. Additionally, as Edelstein (1998) has observed, members of "contaminated communities" are often burdened by social stigmata that accompany their predicaments. In the community surrounding Hanford, which depends strongly on sales of agricultural products to national and international markets, fears of stigmatization and consequent economic loss led some residents to resist the efforts of their self-identified downwinder neighbors to publicize their concerns (D'Antonio 1993). The resulting tensions led to conflicts between downwinders and their neighbors, as well as within the downwinder community itself.

Nuclear History, Memory, and Heritage

Finally, we offer some brief comments regarding the third component of the framework provided by Taylor et al. (2005), nuclear history, memory, and heritage. Historically, the downwinder community can be understood as a product of convergent conditions including decades of inadequately regulated nuclear production and government secrecy, increasing public concerns regarding environmental issues, and growing nuclear anxieties at a dangerous, late Cold War moment. Linked to those conditions is the decline in public perceptions of the legitimacy of the nuclear system—in terms of its scale and lack of transparency, if not its basic premises—that accompanied the disappearance of the Soviet Union as a motivating threat. In this sense, the downwinders provide a case study of a grassroots response to an evolving historical context.

More fundamentally, however, we have argued in this chapter that the downwinders did not simply respond to those historical conditions; rather, their communicative action contributed significantly to the constitution of those conditions. On this reading, communication plays a larger role in shaping history. Historical circumstances provided an exigence, a set of constraints, and a set of resources for the downwinders' communicative action, but their responses also transformed those circumstances, reflexively. Two important outcomes were a set of environmental and health hazards, and a set of democratic failures, constituted and articulated as social problems through the downwinder narrative. Thus as a record of social and political events during the late and post–Cold War periods, history can recognize the downwinder community as both a symptom and a collective agent.

The downwinders have also influenced the distinct but related domains of nuclear memory and heritage.[16] As a result of their engagement, individual and collective memories of the nuclear age are now more variegated. Themes of national unity, scientific and technological progress, and triumphant successes in World War II and the Cold War (see Krupar and Depoe, in this volume) are complicated by competing memories of miscalculation, misrepresentation, and technocratic myopia. Similarly, Hanford's heritage discourse, formerly dominated by the "Cold-warriors" and technological enthusiasts who worked within the reservation's boundaries, now encompasses a broader range of narratives representing the more ambiguous experiences of neighboring, affected communities. Those narratives have the potential to affect evolving understandings of the Cold War and its immediate aftermath, and to provide important lessons regarding communication among government officials, technical specialists, and the public.

CONCLUSION: CONSTITUTIVE COMMUNICATION AND THE CONTEMPORARY RISK SOCIETY

In examining the self-organization of the Hanford downwinders as case study in constitutive communication, we have linked concepts of identification, sense-making, narrativity, symbolic convergence, and contemporary risk society. We believe that this integration of perspectives drawn from rhetoric, narrative theory, organizational and group communication, and sociology can provide a useful model for further studies of communication as a constitutive, social process. Here we have applied that model to examine how the downwinder narrative has simultaneously constituted an environment of material risks, an associated risk community, and a set of related problems and responses within the democratic public sphere.

Through that narrative the downwinders have challenged a consequential formation of institutional power, sustained by the principles of secrecy and monolithic, technocratic rationality. The downwinders problematized those dual principles of discursive containment by offering plausible, coherent narratives that expanded the prevailing dialogue, and through their use of multiple forms of public expertise. In doing so, they simultaneously responded to and contributed to an extraordinary historical moment. Their challenge has influenced the practices of powerful government institutions, and has continuing relevance to problems of secrecy and technocracy in a democratic society.

NOTES

1. Previous versions of this essay were presented at the Eighth Biennial Conference on Communication and the Environment, Jekyll Island, Ga., June 2005; at the Society for Social Studies of Science annual conference, Vancouver, B.C., November 2006; and at the National Communication Association annual conference, San Antonio, Tex., November 2006. The authors wish to thank Victoria J. Gallagher, David R. Seibold, Bryan C. Taylor, Kenneth Zagacki, and a number of anonymous reviewers for their comments on various drafts. We also thank the College of Humanities and Social Sciences at North Carolina State University for support of this essay through a scholarly project award.

2. Gallagher (1993), IPPNW and IEER (1991), Masco (2004), and a recent analysis by the National Academy of Science (2003) provide support for this expansive claim.

3. According to Gephart (2003), during its operational lifetime Hanford produced 74 tons of plutonium, 65 percent of the U.S. supply.

4. A parallel disruption exists with respect to the compact between the U.S. federal government and Hanford's Tribal neighbors, who claim a direct government-to-government relationship. Although Tribal members do not typically identify themselves as downwinders, and tend to distance themselves from non-Tribal negotiations with the U.S. federal government, in some respects the two communities are aligned.

5. In 1983 the Department of Energy chose Hanford as one of nine sites for consideration as an underground repository for high-level nuclear wastes, evoking strong popular opposition from the state of Washington. In 1987 the U.S. Congress limited the repository studies to what would become the Yucca Mountain site in Nevada (Gephart 2003). The future of the Yucca Mountain project remains uncertain.

6. That particular release, which took place December 2–3, 1949, represents only a small fraction of the historical total. Makhijani, Ruttenber, Kennedy, and Clapp (2000, 269) estimate that "During the first two decades of plutonium reprocessing, Hanford released approximately 700,000 curies . . . of iodine-131 into the atmosphere." Geiger and Rush (1992, 24) raise concerns about "the possible exposure of up to 13,000 children" in the region surrounding Hanford, some of them to very high levels of radiation. For additional background, see www.doh.wa.gov/Hanford/publications/history/release.html#Estimates.

7. The incident involved the reprocessing of irradiated reactor fuel before the fission products had decayed to a more tolerable level; that is, while the fuel was still metaphorically "green."

8. The HEDR study began in October 1987, funded by the Department of Energy, and examined air releases as well as releases to the Columbia River. After expressions of concern regarding the need for greater external oversight, funding for the study was transferred to the Centers for Disease Control and Prevention (CDC) in

1992. A summary of results was published in April 1994 (Hanford Environmental Dose Reconstruction Project 1994). The HTDS, a CDC project from the outset, began in 1988 and issued a controversial draft final report in January 1999. A review of that report by the National Academy of Science (2000) validated a number of methodological and management concerns expressed by downwinders (see Kinsella 2001). The final HTDS report was issued in June 2002, concluding that "there was no evidence of a statistically significant association between estimated thyroid radiation dose from Hanford and the cumulative incidence of any of the fourteen primary [thyroid disease] outcomes" studied (Davis, Kopecky, and Hamilton 2002, xlvi). The report did acknowledge, ambiguously, that "Although no statistically significant dose-response was found for any of the disease outcomes in this study, many study participants had thyroid disease" (liii). The report negotiated the difficult tension between scientific rigor and lived experience by stating cautiously that "These findings do not definitively rule out the possibility that Hanford radiation exposures are associated with an increase in one or more of the outcomes under investigation. However, it does mean that if such associations exist, they were likely too small to detect using the best epidemiologic methods available" (liv).

9. Although many downwinders viewed Ruttenbur as unsympathetic to their interests, he has made important contributions to the critique and improvement of epidemiological studies of radiation effects (cf. Makhijani, Ruttenbur, Kennedy, and Clapp 2000). The tension between Ruttenbur and the downwinders reflects a more general distrust of scientific approaches and government affiliations.

10. For details on health-effects issues at the Fernald site, see Hamilton, this volume.

11. *Dead Reckoning* provides a critical review of 124 published epidemiologic studies of workers within the nuclear weapons production complex, conducted or sponsored by the Department of Energy and its predecessor agencies. The first of the authors' fifteen summary recommendations is that "the involvement of the Department of Energy in the supervision of epidemiologic research activities on its workforce and on the health and environmental effects on surrounding communities should be ended completely and definitively" and replaced by an independent research program (Geiger and Rush 1992, 13).

12. Geiger and Rush (1992, 57) report that Till later testified that his "group's work met with unfailing public skepticism once the funding source became known."

13. Gephart (2003) reports that one component of the combined suit sought $100 billion in payments for alleged damages.

14. This incident accounts for only a fraction of the high-level liquid waste that has leaked from Hanford's 177 underground storage tanks. The total amount leaked could be as much as 1.5 million gallons, and an additional 120 million gallons has been discharged into the ground "after some radionuclide removal." The 53 million gallons now in the tanks is approximately 10 percent of the high-level liquid waste produced; "the other 90 percent was treated, discharged to the ground, leaked to the soil, or evaporated" (Gephart 2003, 5.7; National Research Council 2001).

15. Bazerman (2001) argues that similar issues of information management and interpretative authority were present as early as the 1950s in public discourse surrounding the health effects of nuclear weapons testing.

16. See Taylor et al. (2005) and Krupar and Depoe, this volume, regarding the distinctions among history, memory, and heritage.

BIBLIOGRAPHY

Ball, Howard. 1986. *Justice downwind: America's atomic testing program in the 1950s*. New York: Oxford University Press.

Bazerman, Charles. 2001. Nuclear information: One rhetorical moment in the construction of the nuclear age. *Written Communication* 18: 259–295.

Beck, Ulrich. 1992. *Risk society: Towards a new modernity*. Thousand Oaks, Calif.: Sage.

———. 1994. *Ecological enlightenment: Essays on the politics of the risk society*. Atlantic Highlands, N.J.: Humanities Press.

———. 1995. *Ecological politics in an age of risk*. London: Polity.

Bormann, Ernest G., John F. Cragan, and Donald C. Shields. 1996. An expansion of the rhetorical vision component of the symbolic convergence theory: The Cold War paradigm case. *Communication Monographs* 63: 1–28.

———. 2001. Three decades of developing, grounding, and using symbolic convergence theory. *Communication Yearbook* 25: 271–313.

Brown, Phil. 1987. Popular epidemiology: Community response to toxic waste-induced illness in Woburn, Massachusetts. *Science, Technology, and Human Values* 12: 78–85.

Brown, Phil, and Edwin J. Mikkelsen. 1990. *No safe place: Toxic waste, leukemia, and community action*. Berkeley: University of California Press.

Burke, Kenneth. 1969. *A rhetoric of motives*. Berkeley: University of California Press.

Caulfield, Catherine. 1989. *Multiple exposures: Chronicles of the radiation age*. New York: Harper and Row.

Center for Science in the Public Interest. 2006. *Ensuring independence and objectivity at the National Academies*. Washington, D.C.

Cheney, George. 1983a. The rhetoric of identification and the study of organizational communication. *Quarterly Journal of Speech* 69: 143–158.

———. 1983b. On the various and changing meanings of organizational membership: A field study of organizational identification. *Communication Monographs* 50: 324–362.

———. 1991. *Rhetoric in an organizational society: Managing multiple identities*. Columbia: University of South Carolina Press.

Clapp, Richard W. 2002. Popular epidemiology in three contaminated communities. *Annals of the American Academy of Political and Social Science* 584: 35–46.

Cooren, François. 2000. *The organizing property of communication*. Amsterdam: John Benjamins.

Cox, Robert. 2006. *Environmental communication and the public sphere*. Thousand Oaks, Calif.: Sage.

Dalton, R. J., P. Garb, N. Lovrich, J. C. Pierce, and J. M. Whiteley. 1999. *Critical masses: Citizens, nuclear weapons production, and environmental destruction in the United States and Russia*. Cambridge, Mass.: MIT Press.

D'Antonio, Michael. 1993. *Atomic harvest: Hanford and the lethal toll of America's nuclear arsenal*. New York: Crown.

Davis, Scott, Kenneth J. Kopecky, and Thomas E. Hamilton. 2002. *Hanford Thyroid Disease Study final report*. Seattle: Fred Hutchinson Cancer Research Center.

Douglas, Mary. 1966. *Purity and danger: An analysis of the concepts of pollution and taboo*. London: Routledge.

Edelstein, Michael R. 1988. *Contaminated communities: The social and psychological impacts of residential toxic exposure*. Boulder, Colo.: Westview.

Fischer, Frank. 2000. *Citizens, experts, and the environment: The politics of local knowledge*. Durham, N.C.: Duke University Press.

Fisher, Walter R. 1987. *Human communication as narration: Toward a philosophy of reason, value, and action*. Columbia: University of South Carolina Press.

Flyvbjerg, Bent. 1998. *Rationality and power: Democracy and practice*. Chicago: University of Chicago Press.

———. 2001. *Making social science matter: Why social inquiry fails and how it can succeed again*. Cambridge: Cambridge University Press.

———. 2003. *Megaprojects and risk: An anatomy of ambition*. Cambridge: Cambridge University Press.

Fradkin, Philip L. 1989. *Fallout: An American nuclear tragedy*. Tucson: University of Arizona Press.

Gallagher, Carole. 1993. *American ground zero: The secret nuclear war*. Cambridge, Mass.: MIT Press.

Geiger, H. Jack, and David Rush. 1992. *Dead reckoning: A critical review of the Department of Energy's epidemiologic research*. Washington, D.C.: Physicians for Social Responsibility.

Gephart, Roy E. 2003. *Hanford: A conversation about nuclear waste and cleanup*. Columbus, Ohio: Battelle Press.

Gerber, Michele S. 1992. *On the home front: The Cold War legacy of the Hanford nuclear site*. Lincoln: University of Nebraska Press.

Goldberg, Stanley. 1998. General Groves and the Atomic west: The making and the meaning of Hanford. Pp. 39–89 in *The atomic west*, ed. Bruce Hevly and John M. Findlay. Seattle: University of Washington Press.

Greimas, Algirdas J. 1987. *On meaning: Selected writings in semiotic theory*. London: Frances Pinter.

Hacker, Barton C. 1998. "Hotter than a $2 pistol": Fallout, sheep, and the Atomic Energy Commission 1953–1986. Pp. 157–175 in *The atomic west*, ed. Bruce Hevly and John M. Findlay. Seattle: University of Washington Press.

Hales, Peter B. 1997. *Atomic spaces: Living on the Manhattan project*. Urbana: University of Illinois Press.

Hanford Environmental Dose Reconstruction Project (HEDR). 1994. *Summary: Radiation dose estimates from Hanford radioactive material releases to the air and the Columbia River*. Olympia: Washington State Department of Ecology.

Hanford Federal Facility Agreement and Consent Order. 2003. Washington State Department of Ecology, U.S. Environmental Protection Agency, and U.S. Department of Energy. Available at www.hanford.gov.

Hein, Teri. 2003. *Atomic farmgirl: The betrayal of Chief Qualchan, the Appaloosa, and me*. New York: Houghton Mifflin.

Hope, Diane. 2004. The rhetoric of autobiography in women's environmental narratives: Lois Gibbs' *Love canal: My story* and Sandra Steingraber's *Living downstream: An ecologist looks at cancer and the environment*. Pp. 77–97 in *Environmental Communication Yearbook, vol. 1*, ed. S. L. Senecah. Mahwah, N.J.: Erlbaum.

International Physicians for the Prevention of Nuclear War and Institute for Energy and Environmental Research (IPPNW and IEER). 1991. *Radioactive heaven and earth: The health and environmental effects of nuclear weapons testing in, on, and above the earth*. New York: Apex.

Iodine-131 deposition in cattle grazing on north margin of Hanford Engineer Works. 1946. Staff report HW-3-3628. Richland, Wash.: Richland Laboratories.

Kaplan, Louise. 1992. A slow motion emergency: Radiation health effects and Hanford. Unpublished doctoral dissertation, Brandeis University.

———. 2000. Public participation in nuclear facility decisions: Lessons from Hanford. Pp. 67–83 in *Science, technology, and democracy*, ed. Daniel L. Kleinman. Albany: SUNY Press.

Kinsella, William J. 2001. Nuclear boundaries: Material and discursive containment at the Hanford nuclear reservation. *Science as Culture* 10: 163–194.

———. 2002. Problematizing the distinction between expert and lay knowledge. *New Jersey Journal of Communication* 10: 191–207.

———. 2004a. Public expertise: A foundation for citizen participation in energy and environmental decisions. Pp. 83–95 in *Communication in environmental decision making: Advances in theory and practice*, ed. Stephen P. Depoe, John W. Delicath, and Marie-France Aepli Elsenbeer. Albany: SUNY Press.

———. 2004b. Nuclear discourse and nuclear institutions: A theoretical framework and two empirical examples. *Qualitative Research Reports in Communication* 5: 8–14.

———. 2005. One hundred years of nuclear discourse: Four master themes and their implications for environmental communication. Pp. 49–72 in *Environmental Communication Yearbook, vol. 2*, ed. Susan L. Senecah. Mahwah, N.J.: Erlbaum.

Laird, Frank N. 1993. Participatory analysis, democracy, and technological decision making. *Science, Technology, & Human Values* 18: 341–361.

Limerick, Patricia N. 1998. The significance of Hanford in American history. Pp. 53–70 in *Terra Pacifica: People and place in the northwest states and western Canada*, ed. Paul W. Hirt. Pullman: Washington State University Press.

Lodwick, Dora G. 1993. Rocky Flats and the evolution of mistrust. *Research in Social Problems and Public Policy* 5: 149–170.

Loeb, Paul. 1982. *Nuclear culture: Living and working in the world's largest atomic complex*. New York: Coward, McCann, and Geoghegan.

Luhmann, Niklas. 1989. *Ecological communication*. Trans. John Bednarz, Jr. Chicago: University of Chicago Press.

Makhijani, Arjun, Howard Hu, and Katherine Yih, eds. 2000. *Nuclear wastelands: A global guide to nuclear weapons production and its health and environmental effects*. Cambridge, Mass.: MIT Press.

Makhijani, Arjun, A. James Ruttenbur, Ellen Kennedy, and Richard Clapp. 2000. The United States. Pp. 169–284 in *Nuclear wastelands: A global guide to nuclear weapons production and its health and environmental effects*, ed. Arjun Makhijani, Howard Hu, and Katherine Yih. Cambridge, Mass.: MIT Press.

Mancuso, Thomas F., Alice M. Stewart, and G.W. Kneale. 1977. Radiation exposures of Hanford workers dying from cancer and other causes. *Health Physics* 33: 369–384.

Masco, Joseph. 2004. Mutant ecologies: Radioactive life in post–Cold War New Mexico. *Cultural Anthropology* 19: 517–550.

National Academy of Science, Board on Radiation Effects Research. 2000. *Review of the Hanford Thyroid Disease Study Draft Final Report*. Washington, D.C.: National Academies Press.

———. 2003. *Exposure of the American population to radioactive fallout from nuclear weapons tests*. Washington, D.C.: National Academies Press.

National Research Council. 2001. *Science and technology for environmental cleanup at Hanford*. Washington, D.C.: National Academies Press.

Nussbaum, Rudi H., Patricia P. Hoover, Charles M. Grossman, and Fred D. Nussbaum. 2004. Community-based participatory health survey of Hanford, Wash., downwinders: A model for citizen empowerment. *Society and Natural Resources* 17: 547–559.

Pearce, W. Barnett, and Stephen W. Littlejohn. 1997. *Moral conflict: When social worlds collide*. Newbury Park, Calif.: Sage.

Peterson, Tarla R. 1997. *Sharing the Earth: The rhetoric of sustainable development*. Columbia: University of South Carolina Press.

Ratliff, Jeanne N. 1997. Improving environmental advocacy: How the Hanford Education Action League challenged the Department of Energy. *Journal of the Northwest Communication Association* 25: 42–59.

Ratliff, Jeanne N., and Michael Salvador. 1994. Building nuclear communities: The Hanford Education Action League. Paper presented at annual conference, Speech Communication Association, New Orleans (19–22 November). Available at eric.ed.gov.

Rhodes, Richard. 1995. *Dark sun: The making of the hydrogen bomb*. New York: Simon and Schuster.

Richardson, Mary, Joan Sherman, and Michael Gismondi. 1993. *Winning back the words: Confronting experts in an environmental public hearing*. Toronto: Garamond Press.

Roff, Sue R. 1995. *Hotspots: The legacy of Hiroshima and Nagasaki*. New York: Cassell.

Sanger, S. L. 1995. *Working on the bomb: An oral history of WWII Hanford*. Portland, Ore.: Portland State University.

Senecah, Susan L. 2004. The trinity of voice: The role of practical theory in planning and evaluating the effectiveness of environmental participatory processes. Pp. 13–33 in *Communication in environmental decision making: Advances in theory and practice*, ed. Stephen P. Depoe, John W. Delicath, and Marie-France Aepli Elsenbeer. Albany: SUNY Press.

Silverman, M. Joshua. 2000. *No immediate risk: Environmental safety in nuclear weapons production 1942–1985*. Unpublished doctoral dissertation, Carnegie Mellon University.

Smyth, Henry D. 1945. *Atomic energy for military purposes*. Washington, D.C.: U.S. Government Printing Office.

Stewart, John. 1995. *Language as articulate contact: Toward a post-semiotic philosophy of communication*. Albany: SUNY Press.

———. ed. 1996. *Beyond the symbol model: Reflections on the representational view of language*. Albany: SUNY Press.

Taylor, Bryan C. 1997. Shooting downwind: Depicting the radiated body in epidemiology and documentary photography. Pp. 289–328 in *Transgressing discourses: Communication and the voice of the other*, ed. Michael Huspek and Gary Radford. Albany: SUNY Press.

Taylor, Bryan C., and Brian Freer. 2002. Containing the nuclear past: The politics of history and heritage at the Hanford Plutonium Works. *Journal of Organizational Change Management* 15, 563–588.

Taylor, Bryan C., William J. Kinsella, Stephen P. Depoe, and Maribeth S. Metzler. 2005. Nuclear legacies: Communication, controversy, and the U.S. nuclear weapons production complex. Pp. 363–409 in *Communication Yearbook, vol. 29*, ed. Pamela J. Kalbfleisch. Mahwah, N.J.: Erlbaum.

Udall, Stewart L. 1994. *The myths of August: A personal exploration of our tragic Cold War affair with the atom*. New York: Pantheon.

U.S. Department of Energy (USDOE). Office of Environmental Management. 1995. *Closing the circle on the splitting of the atom: The environmental legacy of nuclear weapons production in the United States and what the Department of Energy is doing about it*. Washington, D.C..

U.S. House of Representatives. Committee on Energy and Commerce. 1986. *American nuclear guinea pigs: Three decades of radiation experiments on U.S. citizens*. Washington, D.C.

Walker, Charles A., Leroy C. Gould, and Edward J. Woodhouse, eds. 1983. *Too hot to handle? Social and policy issues in the management of radioactive waste*. New Haven, Conn.: Yale University Press.

Wasserman, Harvey, and Norman Solomon. 1982. *Killing our own: The disaster of America's experience with atomic radiation*. New York: Dell.

Weick, Karl E. 1995. *Sensemaking in organizations*. Thousand Oaks, Calif.: Sage.

Williams, Bruce A., and Albert R. Matheny. 1995. *Democracy, dialogue, and environmental disputes: The contested languages of social regulation*. New Haven, Conn.: Yale University Press.

Williams, S. J., chair. 1948. *Report of the Safety and Industrial Health Advisory Board*. Washington, D.C.: U.S. Atomic Energy Commission.

Wynne, Brian. 1996. May the sheep safely graze? A reflexive view of the expert-lay knowledge divide. Pp. 45–80 in *Risk, environment, and modernity: Toward a new ecology*, ed. Scott Lash, Bronislaw Szerzynski, and Brian Wynne. Newbury Park, Calif.: Sage.

4

Regional Communication and Sense of Place Surrounding the Waste Isolation Pilot Plant

Eric L. Morgan

Isolation—The placement of radioactive waste so that contact between the waste and humans or the environment will be highly unlikely for a chosen period of time.

—U.S. Environmental Protection Agency, *Invitation for comment on the criteria for radioactive waste*, Federal Register 1978, 53262

Of the possible ways to deal with the radioactive waste produced by the nation's nuclear weapons industry, "isolation" has long held the attention of policy makers and producers. The above definition operates on the specific assumption that nuclear waste can, in some way, be removed from contact with humans and the environment. That assumption has led to the construction and operation of the nation's first underground repository for the disposal of transuranic nuclear waste. The Waste Isolation Pilot Plant (WIPP) opened its doors on March 26, 1999, to receive the first shipment of waste to be isolated from humanity and the environment. The isolating takes place 2,150 feet (about four-tenths of a mile) below the surface of the ground, in a 2,000-foot-thick salt bed in a remote corner of southeastern New Mexico.

Whether nuclear waste can actually be isolated has been the source of much debate surrounding the WIPP site. Clearly, digging a very deep hole might remove waste from some human contact, but it only displaces it within the environment. These problems have led to many discussions surrounding "site suitability," or the degree to which any one site has the most desirable characteristics for achieving isolation. Arguments for and against the attractiveness of the site have dominated public discourse surrounding the WIPP project since at least 1979. Indeed, this topic has been institutionalized with the formation of an Environmental Evaluation Group, whose role was and is to provide "technical evaluations on the design, construction, and operation of the WIPP project" (Allen 2001, 3). As a site, WIPP is able to achieve a sufficient degree of suitability because of the ways in which it has been discursively framed as safe, beneficial to the economy, and remotely located. However, this framing coupled with the very function of the plant creates a contradiction. As long as humans must contend with the waste produced from nuclear weapons production (a commonly held duration is 10,000 years, although one engineer told me that we really need to talk in terms of millions of years), this waste will always resonate discursively. It is the resilience of these types of discourses that makes "isolation" impossible precisely because nuclear waste and its associated places exist symbolically. Because numerous discourses surrounding the WIPP site are poignant for many stakeholders, transuranic waste will continue to be a part of the communicative structure of group life for those communities that are thrust into the role of maintaining its isolation from the rest of humanity. The project thrusts residents of New Mexico into the role of "nuclear worker," thereby creating a "highly ambiguous and charged social space" (Masco 2004, 518).

This chapter seeks to explicate the structure of those discourses as experienced by residents of Carlsbad in southeastern New Mexico and other communities surrounding the WIPP site. In doing so, the chapter seeks to contribute further to the burgeoning study of nuclear communication. While many studies (e.g., Kinsella 2001; Taylor 1998) focus on the institutions that sustain the nuclear program within the U.S. Department of Energy (DOE), others focus on the social and cultural understandings that the nuclear condition invokes (e.g., Hevly and Findlay 1998; Kuletz 1998; Taylor 1996). One study that focuses on how remembrances of the DOE's Fernald site contribute to an enduring, but at times contrasting, sense of place for some community members (Barnes-Kloth, Depoe, Hamilton and Lombardo 2001) is particularly relevant to understanding the WIPP site. The current study is similar in that it also investigates the link between sense of place and a nuclear site. However, because WIPP is currently operational the present study focuses less on memory and place and more on the discursive system that invokes and implicates different communicative practices. The Fernald study is also similar in that both seek to advance our theoretical understanding of nuclear communication by refocusing the analytic lens on the structure of discourse produced by those who not only have a stake in these sites, but also live in and around them. This leads to the primary research question that investigates how stakeholders and community members make sense of the WIPP site through communication. In addressing this question, a number of other questions become relevant. For instance, are the tropes around which nuclear discourse tends to circulate in scientific, governmental, and academic circles relevant to the people who live and work around nuclear sites? What sense of place is active for these people? How has that sense changed over time? These are all important questions, not simply for their theoretical value, but because they potentially offer practical understandings of the nuclear situation that other communities might use.

In order to address the above questions, this chapter utilizes a theoretical framework provided by cultural discourse theory (Carbaugh, Milburn, and Gibson 1997; Morgan 2002, 2003). This perspective allows us to understand communicative practices in context thereby enabling an understanding of local effects. The perspective allows the researcher to explore the systemic meanings that make relevant more particular meanings surrounding and grounding human social behavior. The essay presents findings from a study of the WIPP site using ethnographic methods implicated by cultural discourse theory. These findings are presented according to the relevant concepts of the framework, followed by a broader discussion of their implications. First, however, an introduction to the relationship between nuclear waste, communication, place, and the WIPP site is in order.

COMMUNICATION AND THE CONFIGURATION OF NUCLEAR SPACES

One underlying assumption of the present essay is that places are made meaningful through the communication practices that invoke them. As people participate in discourses about places, underlying premises about who can inhabit those places, how we should feel about those places, possible relations within those places, and, most importantly, how people are said to be able to act in places all become relevant (Carbaugh 1996a; Morgan 2002). The nuclear condition in which many places throughout the world exist configures the meaning of places in a multitude of ways. These places have become what Hales (1997) called "atomic spaces." What follows is a brief review of the types of nuclear spaces that occur in the United States, and the types of communication practices that bring these places into discourses that then invoke a range of meanings.

Types of Nuclear Places

For a place to be considered a nuclear place within a communication context, it should in some way be available for invocation within a broader context of nuclear discourse and have nuclear materials present. Taylor (1993a, 1993b) has argued that places associated with the nuclear weapons organization can be considered to be complex sites. These sites comprise both material aspects and contested symbolic aspects. What is excluded in this conceptualization are places in which policy makers and interested stakeholders may debate nuclear topics, although these places are also important to an overall understanding of nuclear discourse. With such a definition in mind, nuclear places can be divided into three types including places of production, places of maintenance and regulation, and places of remediation. These types of places are made meaningful through the communicative practices of various stakeholders.

Places of Nuclear Production

Included in this category of place are those sites that are actively dedicated to the production of nuclear weapons, products such as energy, and other related technologies. Within the context of the United States, there is a distinction between those places that utilize nuclear technology for commercial purposes and those that utilize nuclear technology for military purposes. Both types of sites will be discussed briefly here, although it is important to remember that the WIPP site is associated specifically with the nuclear weapons complex.

Commercial Sites

Commercial sites are mostly dedicated to the production of nuclear energy. According to the U.S. Nuclear Regulatory Commission (NRC), there are 104 nuclear power reactors and another 97 research and test reactors located at 69 different sites in the United States (USNRC 2006). The NRC administers these sites in four separate regions. Region I encompasses the northeastern portion of the United States, Region II encompasses the southern portion of the United States including Puerto Rico and the Virgin Islands, Region III encompasses the Midwest, and Region IV encompasses the western portion of the country including Alaska and Hawaii. In Region I, there are 26 nuclear power reactors and 24 research and test reactors located at 18 different sites. In Region II there are 33 power reactors and 17 research and test reactors again located at 18 different sites. Region III contains 24 power reactors and 16 research and test reactors located at 19 different sites. Region IV contains 21 power reactors and 40 research and test facilities located at 14 different sites. Places of production also include those sites that produce "source materials" such as uranium or thorium. It is worth noting that the 26 uranium mining facilities in the United States are all located in Region IV and most of those are located in the Southwest not very far from the WIPP site. However, approximately 70 percent (140 out of 201) of all reactors are located in the other three regions. The variety of facilities spread across the nation requires that nuclear materials be transported. Because of this, the entire nation is potentially an "atomic space." The NRC's configuration of the entire country as an atomic space bears investigation that is beyond the scope of this essay. The fact that it does so, however, draws our attention to a way of sensing place through the lens of the nuclear.

Military Production Sites

Military production sites are those sites that are in some way dedicated to the creation and maintenance of the nation's nuclear weapons program administered by the Department of Energy. Since the time of the Manhattan Project, which began in 1942 at the behest of President Franklin Roosevelt, a number of facilities have been operational in this context; however, there are now far fewer. Barnes-Kloth et al. (2001, 304) state that there were more than 16 major facilities and "thousands of smaller support facilities" involved in this complex. Currently, there are six major sites still supporting the nuclear weapons industry. Three sites are devoted to research and design including Los Alamos National Laboratory and Sandia National Laboratory, both in New Mexico, and Lawrence Livermore National Laboratory in California. Limited material production occurs now only at the Oak Ridge National Laboratory in Tennessee and the Savannah River Site in South Carolina. Weapons assembly and disassembly take place only at the Pantex plant in Amarillo, Texas. While each of these sites is primarily production oriented, they also must contend with the byproducts of nuclear work in the form of nuclear waste. Most of the byproducts produced by the weapons facilities are stored on-site. On-site storage of nuclear waste creates a different type of place with a different set of attendant communication practices.

The communicative practices associated with the performance of nuclear weapons production have been investigated in a number of important studies. Perhaps the most famous is Carol Cohn's (1987) essay concerning gendered language use and modes of thinking among nuclear defense intellectuals. In this piece, she argues that the defense industry in the latter years of the Cold War was configured by a type of thinking that she labels "technostrategic." Technostrategic thinking arises from the "intertwined and inextricable nature of technological and nuclear strategic thinking" (186). This point becomes important in that Cohn argues that technological innovation fundamentally drives nuclear strategic thinking. This, in turn, circumscribes other ways of thinking about the nuclear weapons complex. It is clear that technostrategic thinking implicates a type of technostrategic communication. This type of communication is used to make sense of nuclear weapons, nuclear weapons production, weapon deployment, and what can only be called weapon-human relationships. The use of technostrategic communication, for instance, functions to construct nuclear weapons as somehow more "clean" than conventional weapons and, amazingly, somehow more humane than conventional weapons.

Places of Nuclear Maintenance

One of the consequences of utilizing nuclear technology is the creation of nuclear byproducts and waste. What to do with the enormous amounts of nuclear waste has long been the subject of political and scientific debate in this country. As stated in the introduction, the general sentiment for the management of nuclear waste suggests that we "isolate" it in nuclear waste repositories. Currently, there is but one operational facility dedicated to isolation of nuclear waste from the nation's nuclear weapons program. This is the WIPP site in New Mexico and the focus of this essay. Other proposals for waste storage are currently being debated, the most popular of which is the development of Yucca Mountain in Nevada as a place to store high-level nuclear waste primarily produced by the nation's nuclear energy plants. Located nearby is the Nevada Test Site. It is no longer a site of nuclear testing, but now serves as a storage facility. Other sites are still being debated, such as the lands of the Goshute American Indians in Utah (Taylor 2003). As argued later, the discursive practices that configure one such place of nuclear maintenance center around the themes of economy and isolation.

Places of Remediation

As discussed in the previous section, there is a great amount of controversy surrounding the storage of nuclear waste. Part of this controversy arises precisely because nuclear waste is so detrimental to the environment. This consequence of nuclear waste has led to the closure of numerous facilities with these spaces now given over to environmental remediation. Three such sites have been investigated from a communication standpoint, thus providing some insight into how senses of these places are constructed through communicative practices. These sites include the Fernald site near Cincinnati, Ohio; the Hanford site near Richland, Washington; and the Rocky Flats plant northwest of Denver. Two of these are discussed here.

Barnes-Kloth et al. (1999) present data gathered from six interviews conducted with people associated with the Fernald site. From these data, they derive five themes that explain differing ways of making sense of this place. They ultimately argue that expressed senses of place vary across individuals according to one's past or current relationship with the site. Thus, sense of place is taken to be a psychological characteristic as opposed to a set of discursive practices that allow people to make a common sense of place. An analysis of the Rocky Flats plant by Taylor and Davis (1999) also provides some insight into how places of remediation are configured through communication. The Rocky Flats plant was, at one time, a major facility at which nuclear warhead components were fabricated. In 1992, the plant's mission necessarily changed to environmental remediation after numerous instances of environmental contamination. Although Taylor and Davis's piece does not focus on sense of place directly, their analysis of the role of ideology in decision making across citizen advisory boards and other expert groups allows us to see how Rocky Flats, as a site of remediation, becomes the focal point of contesting discourses. Thus, a site of remediation is a site of contestation.

In summary, nuclear places can be divided into at least three types including places of nuclear production, places of nuclear maintenance and regulation, and places of remediation. Each type of place is configured by types of communication specific to material function and place. These types of nuclear places are but one way to categorize places according to material function. Other types of organizations could also be presented based on differing criteria. In contrasting nuclear places by material function, however, it is possible to track how situated communicative practices come to configure certain places in certain ways. Indeed, each place is worthy of investigation. The remainder of this essay will explore more specifically one such place in which issues of relevant action and dwelling will be highlighted.

THE SCENE MADE RELEVANT THROUGH DISCOURSE

The federal courthouse in Santa Fe, New Mexico, was the scene of a brewing controversy in 1989. The event was another set of public hearings on the then-proposed Waste Isolation Pilot Project, to be located 270 miles south in the salt beds of southeastern New Mexico. I was sixteen and armed with a sense of righteousness regarding all things environmental. A friend and I showed up, picket signs in hand, ready to have our voices heard concerning this travesty. It seemed so simple in Santa Fe to note the clear moral wrong being exacted upon the people of the state. As I sat and listened to the various arguments as to why the WIPP site was a bad idea, several things became clear. First, this was already a done deal. Second, most people there didn't really care about southeastern New Mexico. Rather, they were more interested in the potential disaster of trucks with TRU-PACT containers[1] barreling down St. Francis Drive in Santa Fe, carrying deadly nuclear waste from that den of secrecy twenty minutes northwest known as Los Alamos. That concern was alleviated with the building of a bypass (NM 599) that has had a tremendous environmental impact upon the development of Santa Fe. The building of the bypass was also successful in removing WIPP from the table of public discourse in the northern part of the state. Third, and perhaps most importantly, I realized that the issues being talked about in Santa Fe were not necessarily those that were being talked about in Carlsbad, Artesia, Hobbs, Lovington, Loving, and Roswell. It dawned on me that the folks I knew down there, most of them family members, had a completely different outlook. What were the issues for them? How were they reacting to the storage of nuclear waste squarely in their backyard? Why didn't anyone down there seem to care?

The experience described above provided the impetus for this project. Over the years, discourse surrounding WIPP has become increasingly less prominent for many New Mexicans, becoming a part of the banal backdrop of everyday life (cf. Taylor et al. 2005). Sites of nuclear activity have long been the purview of powerful institutions such as the Department of Energy, the Department of Defense, and government contractors such as DuPont, General Electric, and Westinghouse. Nevertheless, these sites are still places inhabited by people who create, maintain, and transform their worlds through locally sensible communication practices. If sites of nuclear activity are to be understood, then how they function symbolically for those people must also be understood.

THE WIPP SITE

The WIPP Site has been a focus of public discourse in New Mexico since the project's authorization by Congress in 1980. The site in New Mexico was chosen through the efforts of a state senator after another similar geologic site near Lyons, Kansas, failed to win approval (Applegate 2006). After a number of other acts of legislation were passed, site characterization studies were performed, and certifications were obtained, WIPP opened its doors on March 26, 1999, to receive waste from sites around the country. But what exactly is the WIPP site? According to the Department of Energy, WIPP is "the nation's first operating underground repository for defense-generated transuranic radioactive waste" (USDOE 2003a, 1). It is a site that, after much public debate and commentary, came to actualize the selected method of nuclear waste management by placing it far away from potential human contact. The actual process of isolating low-level nuclear waste is guided by numerous regulations concerning what can be stored, what types of containers can be used, how those containers must be shipped from remote locations, and how waste is stored. Here I address each of these aspects in order to provide the reader with a better sense of what happens at WIPP.

The type of waste that can be stored at WIPP is referred to as "contact-handled transuranic waste" (CH-TRU) (USDOE 2005, 1). Transuranic waste is produced in connection with the use of plutonium in research and in the manufacture of nuclear weapons (USDOE 2003b). Contact-handled transuranic waste is defined as waste with a relatively low surface radiation dose (not greater than 200 millirem per hour [mrem/h]). Specifically, CH-TRU consists primarily of contaminated materials such as clothing, tools, rags, residues, and other types of debris that were previously exposed to greater levels of radiation (USDOE 2003a). The types of containers used to transport CH-TRU are known as Transuranic Packaging Transporter Model II (TRUPACT-II) and the smaller HalfPACT. These containers are designed to hold 55-gallon drums, 85-gallon drums, or waste boxes containing the CH-TRU. The TRUPACT-II holds up to fourteen 55-gallon drums and the Half-PACT holds up to seven drums (USDOE 2003c).

Once the TRUPACT-II and the HalfPACT containers are loaded at remote locations, they are placed onto tractor-trailers and transported to the WIPP site. There, they undergo a security inspection, a radiological survey and a shipping document review (USDOE 2003d). After the inspections, a forklift transports the containers through an airlock into the building. Inside the building, a crane removes the containers and places them on pallets, which are then placed in the waste handling shaft. The pallets then descend 2,150 feet to the repository. Once the materials have been placed in the appropriate disposal room, bags of magnesium oxide are placed on top of and around the containers. Such are the day-to-day operations associated with isolating. Indeed, WIPP represents a marvel of engineering technology, but, as with other repository site debates, the engineering is not the only problem (Freudenberg 2004).

THE THEORETICAL PROBLEM AND FRAMEWORK

The theoretical problem for this chapter is one of how discourses are produced and accessed in order to make sense of places and consequently create a "sense" of those places. In the context of nuclear communication, this sensing or sense-making is accomplished through a number of intersecting discourses. These discourses are articulated with one another in locally sensible ways. Thus, the questions of how this process occurs and how it subsequently structures aspects of human social life become salient. Essentially, this is a problem of "how micro social processes underlie the macro in the social life of nuclear waste" (Taylor 2003, 291).

In order to gain an understanding of these questions, a theoretical framework is necessary to allow consideration of how *places* are brought into discourse and thus made meaningful and relevant. The framework used for the analysis of discourse in this setting is based in a developing perspective that may be labeled *cultural discourse theory* (Carbaugh, Milburn, and Gibson 1997; Morgan 2002, 2003). The basic arguments of this theory relate symbols to other analytical concepts such as codes, discourses, and culture.

Cultural Discourse Theory

Symbols, which can be defined as richly significant communicative devices, exist in their enactment in organized ways (Carbaugh 1995, 1996a, 1996b; Hymes 1962; Philipsen 1992). Furthermore, any use of a symbol invokes a place of significance. I use the term "place" to simultaneously highlight the bounded nature of symbols as well as the situated nature of their use. Any use of a symbolic device creates boundaries of meaning for the users. For example, the term "isolation" used at the beginning of this essay creates a sense of action regarding nuclear waste and simultaneously precludes other actions. Symbols are further rendered significant based upon the context surrounding them. Who utters what, when, and under what conditions is crucial to understanding the meanings, and boundaries of meaning, associated with any symbolic device. Accessing and describing these situated enactments of symbols constitutes the first moment of ethnographic theorizing. The next theoretic move is interpretive and requires a different set of concepts.

To begin, the analyst interested in communication should approach the field with an ear toward discovery and description. In order to accomplish this, the first aspect of the communicative environment to be considered is the *context* in which communication occurs. The importance of context cannot be understated. As many ethnographers of communication and sociolinguists have argued before, communication never takes place outside a context. It is always (and already) situated, making its associated interpretations situated as well.

After a thorough consideration of the context of a communicative event, it is possible to then consider certain *key and potent symbols* (Carbaugh 1996a). Key symbols are those which carry a certain degree of resonance within a particular speech community (Carbaugh 1989; Hastings 1995; Schneider 1976). As key symbols, they serve to anchor certain symbolic clusters, which consist of associated terms. These associated symbols are ones that tend to cooccur with the key symbols and are partially made meaningful within a system of local use. Carbaugh (1996a) argues that when attending to these key symbols in relation to the natural environment, attention is drawn to certain aspects of places over others. Consequently, other aspects of the natural environment are hidden from view. This rather simple point has profound effects when we realize that it is precisely this symbolic process that begins the creation of any orientation to the environment. The task of the present project is to discover how these symbolic processes work to both highlight and hide particular aspects of the natural environment and to determine which particular actions are coded within those discursive formulations.

Recognizing how symbols cluster together to form coherent wholes has provided a useful method of analysis (e.g., Burke 1941; Carbaugh 1989; Lindsley 1999). Very few researchers, however, discuss what it is that organizes these symbolic clusters. In work closely aligned with that of Gerry Philipsen (1992, 1997), Carbaugh (1996a) argues that the concept of *code* is useful in terms of discussing how symbols organize into clusters and how those clusters organize in relation to one another. Carbaugh's (1996a, 45) notion of a code refers to an interpretive statement supplied by the analyst as an attempt to describe the relationships between symbols as they exist within "a larger natural and symbolic system." Considering symbols in this manner will highlight certain types of relationships between symbols such as comparative symbols, contrastive symbols, and ones which exist in an agonistic relationship, as well as other formal relations (Carbaugh 1988/89). Other cultural communicative forms that have already been explicated through ethnographic work include a vacillating form (Carbaugh 1996b; Morgan 2002; Wilkins 1999), a ritual form (Carbaugh 1996b; Katriel and Philipsen 1990), and a dramatic form (Carbaugh 1996b; Morgan 2002; Turner 1981). Understanding the formal organization of symbols allows the analyst to consider rules and premises that may be underlie the communicative system at large. As these rules and premises come into view, one can argue that a certain communicative instance, system, or event is coded in particular ways.

Carbaugh (1996b) has identified five culturally grounded categories of premises that can be analyzed separately across cultural groups. These include premises regarding personhood, action, feeling, relating, and dwelling. It must be understood that none of these exists in isolation from any of the others. Rather, focusing on each one provides a slightly different account of the communication conducted within a community. For purposes of this project, I focus on the premise of dwelling. This is a premise that brings one's attention to the relationship between a person (as a cultural being) and the world. Accordingly, premises for action will also be a concern.

The next level of the framework speaks to a broader classification of communicative practices. Here, Carbaugh (1996a) supplies the term *discourse*, which can be defined as the dynamic process of codes in relation to one another. Just as symbols are organized according to codes, so are codes organized according to discourses. Carbaugh ultimately argues that there are three types of discourses: topical, actional, and affiliative. Topical discourses highlight a particular object, concept or topic coded in specific ways. Understanding a discourse as topical requires a knowledge of how speech is coded differently at different times according to different rules and premises. This alternate coding of speech, however, remains focused on the same topic. Actional discourses highlight the coded nature of actions. Understanding this type of discourse is vitally important in terms of answering the broad research question in that it is necessary to understand the manner in which acting in the environment is coded specifically in the locale under consideration. Affiliative discourses focus on the manner in which coded enactments of communication bring people together to form groups. *Culture*, the final concept discussed in the cultural discourse framework, refers to the circulating system of symbols, codes and discourses as they function to, as Geertz (1973) might say, logically integrate the many different levels of meanings operative in a speech community. Defining culture as such reinforces the definition of the concept as a historically transmitted system of symbols in both form and meaning.

This perspective allows for the examination and interpretation of communication practices of some residents of southeastern New Mexico. By attending to the symbolic constructions of the place generally and the WIPP site specifically, it becomes possible to more fully understand the manner in which communication makes possible certain actions in the land. Ultimately, what is accounted for is how places like WIPP are made sensible to the inhabitants of the region and, in a sense, become naturalized. The isolation of nuclear waste thus becomes a part of the everyday communication patterns of local and regional people.

METHODS OF DATA COLLECTION AND ANALYSIS

This project adopts an ethnographic approach to nuclear communication (Taylor 1996), together with an analysis grounded in cultural discourse theory (Carbaugh, Milburn, and Gibson 1997; Morgan 2002, 2003). The primary unit of analysis is a system of symbolic formulations of the WIPP site. Any communication act that referenced WIPP was of potential concern. These references were not always direct, and they varied across a wide range of contexts. These contexts include naturally occurring conversation recorded in field notes, interviews, and written and online sources. The data were collected over the course of one year in several of the towns of southeastern New Mexico and in Las Cruces, New Mexico. For interviewing purposes, I initially sought participants who had direct connection to the WIPP site, but shortly expanded the scope to include any resident of southeastern New Mexico. This approach was useful as it allowed access to different manifestations of the same topical discourse regarding WIPP. The interview guide was designed to be responsive to the research questions; thus, I asked about WIPP and attitudes toward it. I also asked for any narratives of personal experience with the WIPP site.

An initial review for potential themes was conducted as the corpus of data was compiled. This review yielded two broad topic areas that guided the next analytic move, a key term analysis. The following questions were asked of the data set in order to locate the key symbols underlying the communication produced by the residents of southeastern New Mexico. First, which symbols recur across contexts in talk about WIPP? Second, which symbolic formulations co-occur with the key symbols? Third, is there evidence that suggests these terms are prominent? Evidence beyond recurrence and toward prominence might include how the term functions, or even a willingness for participants to elaborate because they feel it is important for the researcher to understand. As the description of the symbol system proceeded, the interpretive move of establishing cultural premises and rules was performed. This method of generating "codes" that function to organize culturally situated communication follows Philipsen's (1997) work identifying "codes of dignity" and "codes of honor" among some residents of Chicago. Following this stage of analysis, a discussion of a broader function of this topical discourse was possible.

FINDINGS: KEY SYMBOLS AND THEIR ORGANIZATION

Within the data set, there are two key symbolic constructions relevant to the WIPP site and the place more generally. The first is a construction associated with a code of economy. The second is a construction associated with a depiction and coding of the place as "nothing." These symbols function together within a topical discourse of place as a nuclear site, and an affiliative discourse that draws community members together around ways of living in the land.

Code of Economy

Upon initial review of the data set, it became evident that participants would invoke WIPP to the extent that it filled a perceived economic need. Thus, symbolic phrases that included the terms "work," "jobs," "financial effect," "security", and "the mines" were particularly significant. Note the following instance from one twenty-four-year old, lifelong resident of Carlsbad whose sister currently works in the offices of WIPP located in

Carlsbad. This respondent is currently enrolled as an undergraduate student at New Mexico State University:

> Well, WIPP has always been there, but I do remember that when it came to town it was a really big deal. I think 'cause of the financial effect it has on the town. A lot of people work there and depend on it.

The suggestion that "WIPP has always been there" highlights an interesting phenomenon. It suggests that while the site itself has only been open to receive nuclear waste for four years, the discursive processes that allowed for the creation of the site are much older. The idea for achieving the goal of isolation through underground storage dates back to 1957, when the National Academy of Sciences concluded that salt deposits were the most promising option for the disposal of transuranic waste (USDOE 2000). Although the initial report stated that there was a lack of evidence to fully support that claim, a later report from the same committee reasserted the benefits of salt bed deposits for the permanent storage of nuclear waste. At that time, the committee supported a site near Lyons, Kansas. That site was subsequently ruled out based on previous drilling in the area and political opposition. In the fall of 1971 New Mexico State Senator Joe Gant, Jr. learned of the Atomic Energy Commission's rejection of the Lyons site, phoned Congressman Harold Runnels, and purportedly asked, "Why not Carlsbad?" Thus, underground storage of nuclear waste in southeastern New Mexico has been discursively relevant for over thirty years. The economic benefits of WIPP have been foregrounded for much of this time. One consequence of this discourse can be seen in my own personal experience as a child of ten, when several of my friends moved from Santa Fe to Carlsbad as a result of parents seeking jobs at the WIPP site.

Long before WIPP's opening, the place was constructed symbolically as an economic windfall. For example, one participant stated, "WIPP means money. They saw dollar signs early on and no one else wanted it. The place [Carlsbad] would be dead without it." Such discursive constructions of economic benefits suggest that there is public sentiment supportive of the WIPP site. As one participant put it when asked if he felt there was some support for the site, "There really is broad support for it in the town, but others are uneducated about it in the state." This view is further evidenced in another statement from the woman quoted above, who said, "Well, WIPP jobs have really good pay, I mean, everybody wants to get a WIPP job." A fifty-five-year-old woman who was born in Carlsbad and grew up in Artesia, but now lives away from the region, noted that, "You have to remember that the people wanted it. Ever since oil and gas went bust, it's hard to stay in the area." Another participant, a thirty-one-year-old park ranger employed at Carlsbad Caverns National Park, highlighted the economic benefits of the WIPP site this way:

> I mean everyone expected Carlsbad to just boom. In ten years, they were saying that there would be double the population. Everybody was excited about that. I remember growing up when they were talking about putting in a mall! I guess they've done that now, but that's all because of the WIPP.

Not only the effect on the economy, but also the type of job, was described positively. WIPP jobs were government jobs. This situation invokes an associated key term, "security," and was summarized by the park ranger when he noted, "everyone wants to get a WIPP job because they're government jobs. I think that's more secure than other kinds of jobs. I remember my stepdad saying there would be jobs for like twenty years." This resident highlights the security that comes with government jobs although WIPP itself is scheduled to close in the year 2016. Further supporting the relationship between job security and WIPP is a recent quote from the mayor of Carlsbad, Robert Forest, in the *Odessa American*, stating, "[WIPP offers] good jobs, good benefits, and great job security."

As the range of meanings associated with the economic prospect of WIPP came into view, it became clear that the positive valuing of the WIPP jobs stood in direct contrast to that of another type of regional job setting, the mines. The potash mines, as a site of economic activity, were discursively formulated in direct relation to the WIPP site. The following statement from one participant who grew up in Carlsbad highlights this comparison: "Again, growing up, WIPP was always there. People really only worked one of two places. You either worked at WIPP or the mines." The park ranger agreed, saying "Everybody's fathers were working for the mines. My dad and my stepdad worked in the mines. The mines, I don't know, growing up, I guess the mines connected people." The assertion by this participant that the mines connected people points to one potential function that WIPP now plays. It suggests that the mines, at one time, fulfilled a social role for many residents of the area by drawing them together as a community. Thus, the mines functioned symbolically to perform a communal function (Philipsen 1992). The mines were also said to be ambiguous as a job source. One interviewee, a former naval officer, also said of the mines in relation to WIPP, "Well, with the mines being sketchy like they are, I think they just closed one, WIPP is the saving grace of Carlsbad." Prior to beginning this project I had underestimated the extent to which the mines functioned within the construction of the place, as highlighted by the following conversation. While explaining the project to a woman, I mentioned that the economy seemed to be a big issue, especially ever since the oil and gas industry had declined. She replied by saying, "Well now, oil and gas money was never really Carlsbad. Most of oil and gas was Artesia [pause] and Hobbs, and even Lovington some, but Carlsbad was always potash." She went on to tell this story:

> You know Artesia and Carlsbad are both Eddy [pause] um Eddy County. Well, used to be. [pause] You know that Carlsbad's the county seat right? [pause] Well, used to be that Artesia always wanted to get away and they would vote on that, but Carlsbad kept voting them down. I guess they wanted that oil and gas money, but there's always been a huge rivalry between Artesia and Carlsbad.

This story speaks to an interesting characterization of Carlsbad as being bigger, and thus more politically powerful, but also more impoverished. This interpretation was further supported by one participant who noted that he and his friends would go to Artesia " 'cause of the restaurants there [pause] and it's just generally nicer." This contrasting of Carlsbad with Artesia has gone on for many years, since well before WIPP was in the picture. As a continuing contrast, it further helped to construct Carlsbad as a place in need of economic support, and it also created the conditions for another construction of the place.

Code of Nothingness

One particular symbolic phrase was heard across the data set as a way to describe much of Carlsbad and southeastern New Mexico. This recurrence is important, I argue, because it contributes to the reasons why WIPP was sited in New Mexico. The symbolic phrase is grounded by the term "nothing" as a potent and resonant depiction of the place. Its use varied in part according to the participants. For those tangentially related to the WIPP site, the place was often characterized by the term "nothing." "There's nothing out here." "There's a whole lotta nothing." However, the symbolic "nothingness" of the place was also invoked by those related to the WIPP site in different ways, such as "It's fairly desolate out here," or "Well, it's not right in town, kinda away from everything." The following passage illustrates this representation of the area:

> When I was in high school um there was like this thing done on Channel One about WIPP [pause]. We were all like excited because here we are this little nothing of a place and they were doing something on us. They did portray us as a bunch a hicks, though.

The enactments of these symbolic formulations combine to structure a particular discourse surrounding the WIPP site. One way to characterize the symbolic shape of the discourse would be to state that WIPP is the monetary lifeblood of the community as it provides "good jobs" and "economic security" for many people. This theme is heard as particularly relevant because of the ambiguity associated with the continuation of potash mining. The need for economic stability can be heard as difficult because of the construction of the place as a space of nothingness. This construction provides a sufficient motive for placing a highly toxic repository in the region as a mechanism for economic security.

Premises underlying the Discourse

To investigate how this place is being constructed symbolically, one can also posit arguments for the existence of underlying premises or assumptions. Such premises lead to potential explanation for ways of acting in the land. An interpretive move in which certain assumptions are posited and then continually tested against other data is a useful approach to the discovery of premises. This type of understanding can be accomplished by asking of the data what assumptions must underlie certain communicative devices. It is important to remember that these interpretive moves arise out of the actual patterned ways of communicating among discourse users in the community.

When investigating talk about WIPP among some residents of Southeastern New Mexico, it becomes clear that there is a premise of *resource* operative for the production and interpretation (Hymes 1972) of communication. In order for the land to be created as a space of nothingness, there must be a contrastive concept. That concept occurs at the point in which the land might be used to achieve economic gains. In the face of declining potash production and seeing the surrounding oil and gas industry decline, some power brokers in the Carlsbad area determined that the region could be put to use by making it a long-term storage facility for radioactive waste. They knew full well that isolation needed maintenance and this meant jobs and economic security.

Interestingly, the most salient arguments against WIPP were formulated and made resonant according to the same premise. One major concern is that future generations may again want to drill for resources in the area. Doing so constitutes the biggest threat to a breach of the WIPP site and subsequent human exposure to radioactive materials. While there are groups that sustain themselves as an alternative voice to WIPP, their arguments are more forcefully heard in places outside the area. Groups such as Citizens for Alternatives to Radioactive Dumping have argued against the site for many years by adopting many of the available technical arguments. That is, opposition is often enacted by arguing that WIPP is a bad idea because it will not work technically. The data from this study suggest that although this may be a potentially useful strategy, especially if opponents were able to locate a fatal flaw in the design, it will not reshape the parameters of the discourse. Instead, the economic trump card will always take precedence over other concerns. One interesting finding (or rather non-finding) concerns the status of human health risks. There were essentially no interview data concerning any associated health risks, although there is much research being conducted in that area. The results of health effects studies could potentially change the structure of the discourse in profound ways. This possibility, however, remains to be realized.

DISCUSSION

The structure of the economic argument is such that it weathers critique from alternate discourse users. The argument is constructed within a discourse that is made meaningful and resilient according to a set of premises and rules for performance. When invoked, this discourse disallows alternative formulations of the WIPP site. At this moment WIPP has enough symbolic force to ground a topical discourse and construct it in economic terms. That same discourse must sustain any viable alternate critique. Indeed, the discourse of groups opposed to the WIPP site gains relatively small purchase upon the economic monolith relevant within the discourse of some community members. One interesting comment was made by a fifty-five-year-old woman who was born in Carlsbad and raised thirty-five miles north in Artesia when she stated, "I don't know what you want to find out, but you know all the people of Carlsbad wanted WIPP. It doesn't matter what they say up [in Santa Fe], the people want WIPP." The question, of course, becomes why this desire on the part of the people of Carlsbad is so prominent. The answer lies in the resuscitation of a town on the brink of economic failure.

Kinsella (2005) argues that there are four themes underlying much of nuclear communication: mystery, potency, secrecy, and entelechy. He offers these themes as a general framework, arguing that they structure nuclear communication in a particular way. From a cultural discourse perspective they could be considered premises that both are invoked and underlie discourse within the nuclear context. Treating them as such yields an interesting analysis.

According to one interviewee, the WIPP site was sold to the people of the area by concerted efforts to alleviate aspects of the mysterious and the potent. Nevertheless, officials reinforced the premise of secrecy through the discursive construction of super-secret buildings with white-coated scientists walking about wearing security badges. Alleviating the mysterious was accomplished by offering numerous opportunities for the townsfolk to have access to the facility. Mystery was further alleviated through the employment of technical discourses with the purpose of explaining what was actually being stored. In those explanations it wasn't big barrels of glowing liquid that were being stored, but instead, articles that had come into contact with radioactive agents and were thus contaminated. This move to counteract a particular conception and representation of nuclear waste also reduced its perceived potency through the invocation of relatively benign objects. Whorf (1953) has argued that people's behavior around certain dangerous items is guided by their linguistic formulations. For example, while working as a fire insurance adjuster Whorf observed that people would treat an empty gas barrel with less caution than a full one because of the semantic dimensions of the adjective "empty." This term denotes absence of substance and thus absence of danger. In reality, however, an empty gas barrel is more dangerous than a full one because vapors are still present and are more combustible. The same logic may apply to the storage of nuclear materials. Efforts by the Westinghouse corporation to promote WIPP were successful in part because of the linguistic formulation of nuclear waste as a collection of benign objects, such as used tools and clothing.[2] Indeed many participants noted that their initial skepticism was countered when they found out "what they were actually doing down there."

Interestingly, the relative "security" that is invoked through the characterization of nuclear waste as benign can also be heard in statements by those opposed to the site. Consider the following quote from one woman testifying during a recent public hearing about the Department of Energy's reapplication for a permit to operate the facility. "If WIPP is supposed to work, we have to make sure that there's not prohibited items in the site. We need to know what's in the drums. Those are the guarantees that Department of Energy made to the citizens of New Mexico. We need to ensure, EPA needs to ensure, that the wrong stuff isn't going into WIPP; that will challenge the integrity of the site." One can hear in this statement the argument that certain types of nuclear waste are better than others. Because many stakeholders in the conversation about WIPP positively value a certain construction of nuclear waste, further support is evident for the resilience of the prevailing economic discourse.

Kinsella's (2005) identification of entelechy as a theme organizing nuclear communication further informs our understanding of the present case. The term "entelechy" invokes notions of an endpoint. Kinsella notes that, as a narrative device, entelechy provides the summation of the narrative. Thus, in relation to WIPP, the question of entelechy remains a possible area of productive research. Questions of how discourse surrounding WIPP invokes endpoints become particularly salient. Certainly, analyses of nuclear contamination risk mitigation over the next 10,000 years are relevant to the development of this concept within nuclear communication. Entelechy, as it can be identified within the discourse of residents of southeastern New Mexico, invokes the possibilities of a "better economic future" in a historically impoverished place. However, concerns with the distant future are notably absent from this topical discourse. It is perhaps because of the sheer amount of time involved that residents must create meanings of their place that recognize nuclear contamination without necessarily invoking some distant future in which the risk has been mitigated.

It seems, then, that the salience of the economic argument, as it is made relevant by the construction of the place as a space of nothingness, is successful partially by eliminating concerns around the danger of nuclear waste. This is done through a reframing in terms that invoke a more benign substance. Once arguments concerning health risks were adequately addressed, discourses of economy and nothingness were articulated in a powerfully resilient fashion. This articulation occurs in the uniquely situated context of Carlsbad and the surrounding area. Given this context, it is important to note that other communities facing similar situations will construct meaning differently according to potentially different sets of premises.

CONCLUSION: WIPP AND SENSE OF PLACE

At this point, the data suggest that the WIPP site is constructed as valuable and positive for the region because of the economic benefits associated with it. Underlying this discourse is a premise that constructs the land purely as a resource to be used. As natural resources dry up in the region, the land becomes a space of nothingness. To maintain its status as "useful," the place becomes acceptable for the storage of dangerous waste for the next ten thousand years. It is a powerful symbolic construction that renders a place suitable for sacrifice for ten millennia. That power is achieved through the theme of economy and a sense of the place as nothing. In terms of the code of economy, WIPP is not unique. Kinsella (2001) identifies a similar privileging of economic development by some stakeholders associated with the Hanford, Washington site. In terms of nothingness, WIPP and its environs function like other places in the Southwest that have been constructed as spaces available to be poisoned in the name of security (Kuletz 1998).

Ultimately, WIPP becomes a location in which the interplay between "site" and "place" are negotiated daily (Perlman 1988). "Home," as a reference to Carlsbad and the surrounding areas, becomes a "location organized for the purposes of instrumental rationality" (Taylor 1997, 214). What resolves the negotiation is the interaction of economic benefit and a "space of nothingness" as symbolic resources. Notably absent from this discourse are alternate discourses in the region surrounding Carlsbad Caverns National Park, which is also constructed as important because of its economic benefits to the area

(Foster 2006). However, the park is necessarily a place of "somewhere." That two contrasting depictions of southeastern New Mexico can exist simultaneously without necessarily being pitted against one another is a testament to the force of the code of economy.

These observations become more interesting in light of other work about the legacy of nuclear weapons in this country. Writing about the Manhattan Project, Hales (1997, 5) notes that, "the Project is also disappearing from the consciousness, and the conscience, of our cultural and moral life. We are in danger of forgetting. But if we do, it will be a peculiar forgetting, the sort that comes when things become so deeply imbedded in everyday discourse, and yet are so painful, ambiguous, and complex that they seem a part of the ground upon which we stand to survey the world and measure our place within it." The evidence from the current project suggests that much the same is happening with the WIPP site. The degree to which WIPP exists within an everyday discourse can be heard in a statement from a young woman born and raised in Carlsbad, who said "I really don't think WIPP is all that big a deal, especially, I mean in terms of waste. I mean, not like it is to the rest of the world. It's just kind of a norm here."

The WIPP site is more than a nuclear waste repository. It is more than a sacrificial zone. It is a site and scene of active discourse. Within that discourse, meanings about place and environment are made relevant for many people who find themselves contending daily with a nuclear condition. As discourse about the WIPP site becomes increasingly mundane, it will become more a part of the structure of everyday life for many residents of southeastern New Mexico. Thus it is important to understand the shape of that discourse, so that as future consequences of this particular practice become clearer, it can be possible to engage in a viable and sensible conversation.

NOTES

1. An explanation regarding these containers appears in a later section of this essay.
2. See Cohn (1987), and Taylor, in this volume, regarding discourses of nuclear "domestication."

BIBLIOGRAPHY

Allen, L. E. 2001. *Evaluation of proposed panel closure modifications at WIPP*. Carlsbad, N.M.: Environmental Evaluation Group.

Applegate, David. 2006. *Updated status of WIPP repository (1-3-01)*. Alexandria, Va.: Government Affairs Program, American Geological Institute. Available from www.agiweb.org/gap/legis106/wipp.html.

Barnes-Kloth, Rhonda, Stephen Depoe, Jennifer Hamilton, and Amy Lombardo. 2001. Memories of Fernald: Defining a "sense of place" through personal narrative. Pp. 303–317 in *Proceedings of the fifth biannual conference on communication and the environment*, ed. Brant Short and Dayle Hardy-Short. Flagstaff: Northern Arizona University.

Burke, K. 1941. *The philosophy of literary form: Studies in symbolic action.* Baton Rouge: Louisiana State University Press.

Carbaugh, Donal. 1988/89. The critical voice in ethnography of communication research. *Research on Language and Social Interaction* 23: 261–281.

———. 1989. *Talking American: Cultural discourses on Donahue*. Norwood, N.J.: Ablex.

———. 1995. The ethnographic tradition of Philipsen and his associates. Pp. 269–297 in *Watershed research traditions in human communication theory*, ed. Donald P. Cushman and Branislav Kovaãiç. Albany: SUNY Press.

———. 1996a. Naturalizing communication and culture. Pp. 38–57 in *The symbolic earth: Discourse and our creation of the environment*, ed. James G. Cantrill and Christine L. Oravec. Lexington: University Press of Kentucky.

———. 1996b. *Situating selves: The communication of identity in American scenes*. Albany: SUNY Press.

———. 1999. "Just listen": "Listening" and landscape among the Blackfeet. *Western Journal of Communication* 63: 250–270.

Carbaugh, Donal, T. Milburn and T. Gibson. 1997. Communication as cultural pragmatic action: A Puerto Rican "dinner" and a private college "discussion." Pp. 1–24 in *Emerging theories of human communication,* ed. B. Kovacic. Albany: The SUNY Press.

Cohn, C. 1987. Sex and death in the rational world of defense intellectuals. *Signs* 12: 687–728.

Federal Register 1978. Invitation for Comment on the criteria for radioactive waste. 43:53262, Washington, D.C.:GPO.

Foster, R. 2006. Communication and sense of place surrounding Carlsbad Caverns National Park, White Sands National Monument, and Guadalupe Mountains National Park. M.A. Thesis. New Mexico State University.

Freudenberg, W. 2004. Can we learn from failure?: Examining U.S. experiences with nuclear repository siting. *Journal of Risk Research*. 7: 153–169.

Geertz, C. 1973. *The interpretation of cultures.* New York: Bantam Books.

Hales, Peter B. 1997. *Atomic spaces: Living on the Manhattan project.* Urbana: University of Illinois Press.

Hastings, S. O. 1995. Social drama as a site for the communal construction and management of Asian Indian "stranger" identity. *Research on Language and Social Interaction.* 34: 309–35.

Hevly, Bruce, and John M. Findlay. 1998. *The atomic West*. Seattle: University of Washington Press.

Hymes, Dell. 1962. The ethnography of speaking. Pp. 13–53 in *Anthropology and Human Behavior,* ed. T. Gladwin and W. Sturtevant. Washington, D.C.: Anthropological Society of Washington.

———. 1972. Models of the interaction of language and social life. Pp. 35–71 in *Directions in Sociolinguistics: The ethnography of communication,* ed. J. Gumperz and Dell Hymes. New York: Holt, Rinehart, and Winston.

Katriel, Tamar and Gerry Philipsen. 1981. "What we need is communication": "Communication" as a cultural category in some American speech. *Communication Monographs* 48: 301–317.

Kinsella, William J. 2001. Nuclear boundaries: Material and discursive containment at the Hanford nuclear reservation. *Science as Culture* 10: 163–194.

———. 2005. One hundred years of nuclear discourse: Four master themes and their implications for environmental communication. Pp. 49–72 in *The environmental communication yearbook, vol. 2*, ed. Susan L. Senecah. Mahwah, N.J.: Lawrence Erlbaum Associates.

Kuletz, Valerie. 1998. *The tainted desert: Environmental and social ruin in the American West.* New York: Routledge.

Lindsley, S. 1999. Communication and "the Mexican way": Stability and trust as core symbols in maquiladoras. *Western Journal of Communication* 63: 1–32.

Masco, Joseph. 2004. Mutant ecologies: Radioactive life in post–Cold War New Mexico. *Cultural anthropology* 19: 517–50.

Morgan, Eric. 2002. Communicating environment: Cultural discourses of place in the Pioneer Valley of Western Massachusetts. Ph.D. dissertation. University of Massachusetts, Amherst. Abstract in *Dissertation Abstracts International*, 63: 1A.

———. 2003. Discourses of water: A framework for the study of environmental communication. *Applied Environmental Education and Communication* 2: 153–159.

Perlman, M. 1988. *Imaginal memory and the place of Hiroshima.* Albany: SUNY Press.

Philipsen, Gerry. 1992. *Speaking culturally: Explorations in social communication.* Albany: SUNY Press.

———. 1997. *Developing communication theories.* Albany: SUNY Press.

Schneider, D. M. 1976. Notes toward a theory of culture. Pp. 197–220 in *Meaning in anthropology,* ed. Keith H. Basso and H. A. Selby. Albuquerque: University of New Mexico Press.

Taylor, Bryan C. 1993a. Register of the repressed: Women's voice and body in the nuclear weapons organization. *Quarterly Journal of Speech* 79: 267–285.

———. 1993b. Fat Man and Little Boy: Cinematic representations of interest in the nuclear weapons organization. *Critical Studies in Mass Communication* 10: 367–394.

———. 1996. Make bomb, save world: Reflections on dialogic nuclear ethnography. *Journal of Contemporary Ethnography* 25: 120–143.

———. 1997. Home zero: Images of home and field in nuclear-cultural studies. *Western Journal of Communication* 61: 209–234.

———. 1998. Nuclear weapons and communication studies: A review essay. *Western Journal of Communication* 62: 300–315.

———. 2003. Nuclear waste and communication studies. *The review of communication* 3: 285–291.

Taylor, B. C., W. J. Kinsella, S. P. Depoe and M. S. Metzler. 2005. Nuclear legacies: communication, controversy, and the U.S. nuclear weapons production complex.

Pp. 363–409 in *Communication Yearbook 29*, ed. Pamela Kalbfleisch. Mahwah, N.J.: Lawrence Erlbaum Associates.

Taylor, Bryan C. and S. Davis. (1999). Environmental communication and nuclear communication studies: The case of Rocky Flats. Pp. 286–299 in *Proceedings of the fifth biennial conference on communication and environment,* ed. C. Brant Short and Dayle Hardy-Short. Flagstaff: Northern Arizona University Press.

Turner, Victor. 1981. *The drums of affliction: A study of religious processes among the Ndembu of Zambia.* Ithaca, N.Y.: Cornell University Press.

U.S. Department of Energy (USDOE). 2000. *Pioneering nuclear waste disposal* CAO0-00-3124. Carlsbad: U.S. Department of Energy Carlsbad Area Office.

———. 2003a. *Why WIPP?* Carlsbad: U.S. Department of Energy Carlsbad Field Office.

———. 2003b. *What is plutonium?* Carlsbad: U.S. Department of Energy Carlsbad Field Office.

———. 2003c. *Transuranic waste transportation containers.* Carlsbad: U.S. Department of Energy Carlsbad Field Office.

———. 2003d. *Step-by-step guide for waste handling at WIPP.* Carlsbad: U.S. Department of Energy Carlsbad Field Office.

———. 2005. *Contact-handled transuranic waste acceptance criteria for the Waste Isolation Pilot Plant.* Carlsbad: U.S. Department of Energy Carlsbad Field Office.

U.S. Nuclear Regulatory Commission (USNRC). 2006. *NRC: Nuclear reactors.* [online]. Available from www.nrc.gov/reactors.html.

Whorf, Benjamin Lee. 1953. *Language, thought, and reality: Selected writings of Benjamin Lee Whorf,* ed. J.B. Carroll. Cambridge, Mass.: MIT Press.

Wilkins, R. 1999. "Asia" (matter-of-fact) communication: A Finnish cultural term for talk in educational scenes. Ph.D. dissertation. University of Massachusetts, Amherst, 1999. Abstract in *Dissertation Abstracts International* 60:1397.

Part Two

ORGANIZING THE PAST, PRESENT, AND FUTURE OF NUCLEAR WEAPONS PRODUCTION

5

Cold War Triumphant

The Rhetorical Uses of History, Memory, and Heritage Preservation within the Department of Energy's Nuclear Weapons Complex

Jason N. Krupar and Stephen P. Depoe[1]

A crackling overhead voice initiates a countdown to zero as tourists sit on benches in a room designed to resemble a concrete bunker. Looking through portholes in the simulated bunker walls, visitors await the expected explosion. When the voice reaches zero, the benches shake, hidden speakers blast low frequencies at tourists to imitate the explosion's shock wave, and overhead lights flare on as an atmospheric nuclear bomb test consumes the movie screen before visitors. This faux bunker and imitation bomb test, shown multiple times daily, reside within the Atomic Testing Museum (ATM). Appropriately, the bunker and its sensory overload experience are labeled the "Ground Zero Theater."

The Atomic Testing Museum officially opened February 20, 2005, in Las Vegas, Nevada, approximately sixty-five miles southeast of the Nevada Test Site. The museum attempts to preserve and present the history of America's nuclear weapons testing program. The facility attracts not only national support and criticism, but also federal monies (Rothstein 2005). Such federal assistance for atomic preservation departs from the norm. Cold War–era nuclear weapons complex relics checker the American landscape. Built and operated out of public sight, these facilities are now in danger of being physically erased from history by the U.S. Department of Energy (DOE). Public inattention permits the federal government to environmentally rehabilitate former weapons sites without implementing historic preservation programs.

While local boosters, federal government officials, and tourists celebrated the opening of the Atomic Testing Museum in 2005, community leaders and preservationists lamented the indifference exhibited by DOE personnel at two former nuclear fabrication centers. Fernald, a uranium refinery in southwest Ohio, and Rocky Flats, a plutonium trigger factory near Denver, Colorado, approached the end of environmental cleanup operations in 2006.[2] Local organizations and state officials tried repeatedly to persuade the Energy Department to construct and maintain museums/multi-use centers that would preserve the histories of the sites. DOE leaders have generally resisted these preservation efforts, even as they moved ahead with remediation activities. The absence of historical information at Fernald and Rocky Flats post-cleanup leaves local populations uninformed about the multiple health and environmental legacies of nuclear production. Unlike the Atomic Testing Museum, proposed centers at Fernald and Rocky Flats would be built on-site or in close proximity (Fernald Citizens Advisory Board 2002, West et al. 2003). Even after the DOE declares Fernald and Rocky Flats cleaned up, the lingering radiation at both locations may constitute a potential hazard to local residents. Preservation and education functions highlighted in the center concepts envisioned by local stakeholders could enhance public awareness of these risks.

Employees working at Fernald, Rocky Flats, and other sites within the weapons complex served on one of the front-lines of the Cold War. Federal policies implemented by the U.S. Atomic Energy Commission (AEC) in the 1950s dictated the dispersal, duplication, and expansion of America's nuclear manufacturing system (History Associates Inc. 1987, USDOE 1997). The perceived Soviet bomber threat convinced AEC leaders to make decisions that protected production continuity. Like their American counterparts, the Soviets were thought to have targeted each known fabrication, laboratory, and testing center in the system. Therefore, site personnel believed themselves to be working under the possibility of atomic annihilation.

During the Cold War period, the histories and the experiences of plant personnel remained cordoned off from public scrutiny. Although local populations initially supported the presence of nuclear manufacturing installations, attitudes changed when the health/environmental costs became publicized. The DOE transitioned from nuclear weapons fabrication to environmental remediation at multiple Cold War sites by the 1990s (USDOE 1997). While DOE administrators expressed commitment to plant cleanup, historical preservation failed to achieve official support. Consequently, the histories and landscapes of sites such as Fernald and Rocky Flats are being scrubbed away. Institutional memories, personal remembrances, and physical artifacts are all in danger of being removed from not only the local sites, but from the national conversation about the legacy of nuclear weapons production. DOE officials, through their actions, are replacing the "atomic spaces" (Hales 1997) of the Cold War nuclear arsenal with a network of historical black holes dispersed throughout the country. When the federal government does support a local preservation initiative, as at the Atomic Testing Museum, such support has resulted in a triumphant display of American nuclear deterrence and policies, with little commentary on the societal costs (Brooks 2005).

In this chapter, we review relevant research in collective memory, nuclear communication criticism, and Cold War studies. Next, we examine the politics of heritage preservation efforts at the Nevada Test Site, Fernald, and Rocky Flats. The three examples demonstrate a range of government responses to local demands for historic programming related to nuclear weapons production and testing. We argue that federal assistance has depended on the willingness of community leaders and organizations to embrace only the official Cold War chauvinism and not a more holistic/encompassing version of historical events. Finally, we analyze a web site related to Rocky Flats that provides an example of critical history that counters sanitized institutional accounts of the site's environmental legacy.

COLLECTIVE MEMORY, NUCLEAR COMMUNICATION CRITICISM, AND COLD WAR TRIUMPHALISM

This chapter is grounded in previous research in three areas. First, we draw from the growing body of work in *communication and collective memory studies*. We start with the traditional delineation between history, memory, and heritage (Taylor, Kinsella, Depoe, and Metzler 2005). As a generic term, history refers both to past events and to efforts of individuals and groups to account for and interpret those events. We view history as rhetorical in at least three senses. First, academic, professional, or cultural historians employ rhetoric when making claims and constructing narratives based on the selection, arrangement, and presentation of textual and material fragments that are by nature polysemic (Carpenter 1995; Clark and McKerrow 1998; Gronbeck 1998; Hasian 1999). Second, political advocates who make arguments about the present and future often draw upon selective accounts and understandings of the past as a rich source of rhetorical invention, a practice termed "ideological history" (Depoe 1994; McGee 1977). Third, history can be a source of critique and liberation when advocates question official accounts of the past through a process of rediscovering lost or silenced voices or perspectives (Cox 1990).

Memory involves the attempt to recount and give meaning to a past event or experience. While memory begins at the personal level, when individuals try to recall what happened to them at particular points in time, it also operates at the group level to the extent that one's identity and systems of meaning are tied to one's sense of place, community, or culture. This shared dimension of remembering is referred to as collective memory (Zelizer 1995).

Researchers have examined how the collective memory of a culture is shaped by the public historical narratives produced by dominant and vernacular voices within that culture (Bodnar 1992; Dickinson 1997; Hasian 2001a, 2001b, 2004, 2005; Phillips 2004). Scholars have analyzed the production of public historical narratives, or "public memory," within a wide variety of cultural artifacts such as memorial sites (Blair, Jeppesen, and Pucci 1991; Gallagher 1995; Jorgensen-Earp and Lanzilotti 1998), commemorative statues (Mandziuk 2003), and motion pictures (Ehrenhaus 2001; Hasian 2001b; Hasian and Carlson 2000), as well as within more conventional political discourse (Bruner 2000; Parry-Giles and Parry-Giles 2000). Stephen Browne (1993, 1995, 1999) has argued that constructions of public memory are profoundly rhetorical, including dimensions of performance, textuality, and contested meaning. Bruner (2000) coined the term "strategies of remembrance" to identify ways that ideological history is used to promote particular viewpoints concerning identity and collective memory within communities. And, importantly for this study, memory work for an individual or community often involves *forgetting*, the selective erasure or reconstruction of past events or experiences in order to maintain or reconstitute one's identity in the present (Browne 1995; Bruner 2000).

Finally, heritage refers to activities by groups, organizations, or institutions, including but not limited to governmental bodies, to preserve and disseminate a particular narrative account that will maintain their viability, status, or power, especially during periods of social change or upheaval (Lowenthal 1998). Communication scholars have analyzed the strategies of remembrance employed in heritage preservation efforts such as organizational histories (Taylor and Freer, 2002) and museum exhibits (Hasian 2004; Hubbard and Hasian 1998; Katriel 1993, 1994; Prosise 1998). Researchers have also discovered efforts to challenge the heritage preservation efforts of dominant groups through practices such as the erection of "counter-monuments" in Germany as a protest against official government memorials to victims of World War II (Young 1992) and, more directly relevant to this study, the display of alternative exhibits at nuclear science museums (Taylor 1997b). We consider heritage preservation as the efforts by organizations to utilize the rhetorical devices of history and public memory in order to reinforce legitimacy, maintain power, and control uncertainty. Like other examples of public historical narrative, heritage preservation shares the rhetorical characteristics of performance, textuality, and contested meaning.

The second major area of research we utilize is *nuclear weapons and communication criticism*. Communication scholars have spent considerable attention on exploring both the material and symbolic dimensions of nuclear weapons production and deployment, along with the environmental remediation of nuclear weapons production facilities (for more thorough reviews see Taylor 1998a, 2003b; Taylor et al. 2005). In this chapter, we are guided by a number of findings in the literature. First, institutional discourse explaining and justifying the production of nuclear weapons, as well as policies involving their possible use, has been characterized as technical and bureaucratic in nature, featuring acronyms, euphemisms, and domesticated language that obscure meanings and power relations (Kauffman 1989). This phenomenon has been aptly summarized by Schiappa (1989) as the rhetoric of "nukespeak." Kinsella (2005) has extended this line of inquiry by identifying mystery, potency, secrecy, and entelechy as four "master themes" that have emerged in nuclear discourse from the start of the Manhattan Project to the present. And, within the context of the environmental remediation of current and former nuclear weapons production facilities, Kinsella (2001) has also examined and critiqued governmental efforts to "contain" both the material contamination and the symbolic meanings associated with that contamination.

Second, multiple, conflicting understandings have emerged and persisted in public discourse concerning the meaning, significance, and dangers of nuclear weapons production, deployment, and cleanup (Smith 1989; Taylor et al. 2005; Wertsch 1987). Despite government attempts to shield nuclear weapons behind a screen of official secrecy and technical jargon, people have maintained a variety of views concerning the benefits and costs of nuclear weapons and the role they have played in recent history. Divergent perspectives have been expressed in a variety of rhetorical situations and through a variety of communication media, including political protests (Glass 1993; Taylor 1993a), official government and unofficial documentary photography (Taylor 1997c, 2003a), and motion pictures (Foss and Littlejohn 1986; Taylor 1993b). Using Kinsella's (2001) containment metaphor, it may be said that nuclear discourse has included the "leakage" of dissenting views that dispute or undercut official versions. Taylor (1996) has encouraged scholars to adopt a "dialogic focus" in order both to identify the voices within nuclear debates and to problematize their authority to limit and determine the outcome of such debates.

Third, and finally, heritage preservation activities within the nuclear weapons complex have been particular sites of struggle over meanings of America's nuclear past. Scholars have examined government attempts to highlight scientific and military achievements associated with nuclear weapons through museum exhibits and other educational programs, as well as efforts by dissenting groups to offer alternative viewpoints (Taylor 1997b). The 1995 controversy over the contents of a proposed Smithsonian exhibit commemorating the fiftieth anniversary of America's use of the atomic bomb on Hiroshima and Nagaski offers a clear illustration of how fiercely advocates compete to frame the collective memory of key events in nuclear history through the important forum of museum exhibits (Hubbard and Hasian 1998; Prosise 1998; Taylor 1998).

An interesting aspect of nuclear heritage preservation involves how the physical, geographic locations where nuclear weapons production and deployment took place have been defined over time. What Hales (1997) has called America's "atomic spaces" are places within the landscapes of America—and the Soviet Union—that were transformed by government edict from homes and farms to official sites of secret government activity. People were displaced, and lives were changed forever, by the spread of the rival nuclear weapons organizations (Taylor 1997a). As the Cold War ended, a small number of production facilities, test ranges, and missile silos in the United States have been cleaned up or restored as historical markers of the nuclear age (Gusterson 2004; Molella 2003). However, the vast majority of them have been abandoned, many seen as environmental wastelands (Kuletz 1998). Vanderbilt (2002) uses the term "black spaces" to refer to the collection of Cold War relics he visited across America, spaces that once constituted the nuclear arsenal but now exist just out of public sight and memory. So, it seems that nuclear heritage preservation includes forgetting as well as remembering the legacies of weapons production and deployment.

The final area of research consulted for this chapter is the burgeoning area of *Cold War studies*. Historians have explored the causes, meanings, and significance of the Cold War between the United States and the Soviet Union since its inception shortly after the end of World War II. When the Berlin Wall fell in 1989 and the Soviet Union dissolved as a unified political structure two years later, historians and others began to focus their attention on how and why the Cold War "ended," and to draw lessons from the conflict to inform military and foreign policy in the post–Cold War period. Debates raged as to whether or not the Soviet system fell because of inherent economic inefficiencies, poor political leadership, or other geopolitical factors (see for example Boyle 2000; Cox 1998; Gaddis 1997; Hogan 1992; Hunter 1998; Katsiaficas 2001; Kotkin 2001; Suri 2002).

In recent years, a growing number of commentators have argued that the United States "won" the Cold War because American political ideals of freedom and democracy and the mighty engine of capitalism prevailed in a fifty-year global struggle versus inferior Soviet values and economics. According to this view, dubbed "Cold War triumphalism" by Schrecker (2004), the U.S. economy dramatically outperformed its Communist counterpart, contributing, along with the rise of American-inspired freedom movements in Eastern Europe, to the eventual collapse of the Soviet system (Beichman 2000; Gaddis 1997). Of particular interest to this study is the claim that America's military, including its vast system of nuclear weapons as well as its combined force structure, played a significant role in defeating the Soviet Union, even though the two powers never faced each other in direct combat (Nichols 2002).

The triumphalist interpretation of the end of the Cold War has become increasingly popular among American politicians in the wake of the 9/11 attacks, with terrorists identified as the new enemy for the twenty-first century (Schrecker 2004). "Cold War triumphalism" has thus provided a historical rationale for increased military actions abroad, and decreased civil liberties and tolerance of dissent at home, in the post–9/11 period (Robin 2004). The use of the triumphalist frame as ideological history parallels other rhetorical uses of American military history, both before and during the Cold War (Kane 1988). The triumphalist frame has played a prominent role in the development of nuclear heritage preservation activities across the country—including the three case studies that we now present.

THE ATOMIC TESTING MUSEUM: PRESERVING THE BOMB

The Atomic Testing Museum (ATM) owes its existence to a decision made by the AEC in 1950 to operate a continental testing site. AEC personnel detonated the first of several hundred nuclear devices at the Nevada Test Site (NTS) in January, 1951. The NTS served as the nation's primary nuclear testing range for the next forty-one years. Thousands worked directly in the testing industry in southern Nevada, and hundreds more earned income in ancillary services. The mushroom clouds from the Test Site could be seen from Las Vegas, contributing to the casino and tourism industries. Each successful test series, observed from the presumed safety of Las Vegas, reaffirmed American prowess and ultimate victory in a conflict devoid of conventional battlefields.

By 1991, the testing industry faced possible extinction (see McNamara, in this volume). The United States and the Soviet Union agreed to reduce their nuclear arsenals in accordance with the Strategic Arms Reduction Treaty (START). The last U.S. underground test shot occurred September 23, 1992, at the Nevada Test Site. A final moratorium on testing took effect that same year in November. While the economic prosperity provided by the NTS came to an end, the gambling and tourism industries of Las Vegas offset the loss.

The Atomic Testing Museum originated from the Coordination and Information Center. The Nevada Operations Office of the DOE opened the Center in July, 1981. The Center was housed at the time on the Bechtel Nevada industrial complex in North Las Vegas (Kimberly 2005). The DOE employed Bechtel Nevada as a primary contractor at the Test Site. The Coordination and Information Center collected and made available to the public historical documents, records, and data concerning radioactive fallout from U.S. bomb testing.

Shortly after nuclear testing ended at NTS in 1992, efforts began to preserve the history and memories of personnel. Volunteers started collecting artifacts and developing traveling exhibits by 1997. These efforts resulted in the formation of the Nevada Test Site Historical Foundation (NTSHF) one year later. The Historical Foundation aimed to conserve and interpret the history of the Test Site. Shortly after its founding, the NTSHF initiated a campaign to construct a museum devoted to the history of America's testing program.

Over the next several years, the Historical Foundation partnered with the Desert Research Institute (DRI), the DOE's National Nuclear Security Administration (DOE-NNSA) Nevada Site Office, and Bechtel Nevada to design and build the Frank H. Rogers Science and Technology Building (Desert Research Institute 2005). Frank H. Rogers served as the chief operating officer of the Test Site during its first decades of use. DRI operates the Southern Nevada Science Park in Las Vegas and serves as the nonprofit research campus of the Nevada System of Higher Education. The Institute has been extensively involved in cultural resource preservation at the Nevada Test Site.

The Frank H. Rogers building's grand opening in October 2003 brought the DRI, the Historical Foundation, the DOE, and Bechtel Nevada together under one roof. The Institute's Center for Arid Lands Environmental Management and its Frank Rogers Center for Environmental Remediation and Monitoring moved into the 66,000-square-foot facility. In addition, DRI's archeologists and technicians in the Division of Earth and Ecosystem Sciences, who managed the DOE's preservation efforts at NTS, relocated to the building. Bechtel Nevada moved the Coordination and Information Center, rechristened the NNSA Nuclear Testing Archive, into new state-of-the-art quarters within the Rogers building. The NTSHF occupied the ground floor of the structure and made plans to open the Atomic Testing Museum.

Building and museum monies came from a variety of sources. The Rogers facility received \$2.7 million in state appropriations and \$8.4 million in state revenue bonds that would be retired from tenant leases and Institute funds (Desert Research Institute 2005). DRI also secured a \$2 million conventional loan. The Historical Foundation raised funds from federal government and private sources, including money from corporations still active at the Nevada Test Site (Rothstein 2005). Lockheed Martin Nevada Technologies contributed \$100,000 to the Historical Foundation in January 2003. Lockheed Martin was another primary contractor at the NTS at the time of the donation. The company instructed that its grant be used to fund the planning, construction, and installation of exhibits in the museum (Nevada Test Site Historical Foundation 2005a). When the museum opened to the public, the NTSHF received approximately \$2 million in federal grants and \$3 million in grants/pledges from private sources (Nevada Test Site Historical Foundation

2005b, Rothstein 2005). The museum's website acknowledged in 2006 that "nearly 50 percent of the funding for the design and construction of the museum" came from Congressional appropriations secured by Senate Minority Leader, Harry Reid (D-Nevada) (Nevada Test Site Historical Foundation 2006). Such federal largess for heritage preservation remains unheard of within the rest of the nuclear weapons complex.

The federal and private monies permitted the Historical Foundation to hire the Canadian firm Andre & Knowlton Associates to design the permanent 8,000-square-foot gallery. Andre & Knowlton specialized in museum exhibits and previously oversaw projects for the Los Angeles County Natural History Museum and the Hong Kong Museum of History (Nevada Test Site Historical Foundation 2005c). Besides hiring gallery design experts, the Historical Foundation also contracted with Maltbie, Inc. to complete and install the actual exhibits and artifacts in the museum (Nevada Test Site Historical Foundation 2005c). Prior to the Atomic Testing Museum project, Maltbie worked on assignments for the Smithsonian Institute's Behring Hall of Mammals and the Liberty Bell Center in Philadelphia (Nevada Test Site Historical Foundation 2005c). The project team for the Testing Museum included NTSHF members, scientists and administrators from DRI, and DOE officials, along with assistance from the national defense laboratories, particularly the Lawrence Livermore National Laboratory.

This government aid resulted in the opening of the museum in 2005 and the creation of a facility packed with artifacts, photos, and interactive exhibits. The permanent gallery space includes a timeline of the nuclear testing program, a wall filled with atomic memorabilia, and a copy of Albert Einstein's letter urging President Franklin D. Roosevelt to investigate the military potential of the atom. Nearby, the museum possesses another 2,000-square-foot space intended for a changing rotation of displays. The design team wanted the permanent exhibits to be highly interactive for audiences. Displays include touch screens, motion-sensitive plasma television presentations, and audio interviews with former test site employees. The Historical Foundation managed to obtain pieces of the Berlin Wall and the World Trade Towers, bringing the museum's storyline into the post–Cold War period.

Adjacent to the museum spaces, the DOE opened the Dina Titus Public Reading Room. The reading room provides access to the 310,000 documents that make up the Nuclear Testing Archive. Dina Titus (D-Las Vegas) served in 2005 as the Nevada State Senate Minority Leader (Rogers 2005), and participated as a member of the museum board of directors (Lake 2005). In addition, Titus taught at the University of Nevada-Las Vegas and wrote critically about the nation's testing program (Titus 2001). She also appears in the video played in the Ground Zero Theater.

The Atomic Testing Museum's opening was not without controversy. Critics, local, regional and national, attacked the museum's content and story presentation. Downwinders Opposed to Nuclear Testing (DONT) issued a statement on February 18, 2005, labeling the museum "revisionist history" in an effort to appropriate terminology usually used by pronuclear advocates (Manning 2005). Thirty-six protestors arrived on the day of the opening. "We really wanted to give voice to some of the missing pieces at the museum. At the very least, history is much more complicated than what is shown," stated Zachary Moon, program and development director of Nevada Desert Experiences, an anti-testing organization that held vigils at the Test Site gates for decades (Manning 2005). Moon helped organize the opening-day protest.

Downwinders, particularly, lashed out at the museum. Historically, downwinders were people who lived in off-site areas, windward of the Test Site where fallout from the demonstration shots fell. Over the years, downwinders filed multiple lawsuits to secure federal compensation for the health effects the radioactive fallout caused in their lives (Taylor 1997c). The perceived absence of information on the downwinders drew immediate negative attention to the Atomic Testing Museum.

In his review of the museum for the *New York Times*, Edward Rothstein wrote, "The history of testing, as told here, is largely the history of its justification. Problems and issues are noted . . . but such issues are mentioned and put aside, to get on with the main story" (Rothstein 2005). Downwinders from across the West denounced the museum and its opening. St. George, Utah, resident and downwinder Michelle Thomas compared the Atomic Testing Museum to a museum "that historically documents the Holocaust but leaves out the stories of its victims" (Downwinders Charge 2005). Other critics wanted to know why taxpayer dollars were spent on a facility that in their opinion was "a walk down memory lane for testing enthusiasts" (Dickson 2005; Downwinders Charge 2005). Little in the museum inspired critics and downwinders in the ATM's ability to "consolidate and preserve the full, complex and vexing history of atomic testing" (Dickson 2005). Mary Dickson, a downwinder and area writer, summarized the perspective of ATM opponents. She stated in the *Salt Lake Tribune*, "the museum cannot be considered complete until it provides a full accounting of the very heavy price Americans paid for our nation's four-decade nuclear testing program. Otherwise, it is nothing more than a propaganda vehicle" (Dickson 2005).

Museum proponents objected strongly to the criticisms that the ATM ignored the downwinders and their experiences. Museum director Bill Johnson told the *Reno News and Review* in March 2005 that this criticism "is not the case" (Myers 2005). He noted that in the Ground Zero Theater, following the simulated bomb blast, the real feature of the venue began with a video presentation that discussed the history of the Test Site and its effects. Johnson further claimed that in the videotape "the downwinders are portrayed prominently" (Myers 2005). However, *Reno News and Review* reporter Dennis Myers noted that a review of the movie dialogue, transcribed and provided by Johnson, revealed the absence of any downwinders in the production (Myers 2005). Two of the video participants were former test site workers, and the third taped individual was Dina Titus.

In fact, before the downwinder controversy, Johnson went on record with the *Christian Science Monitor* stating, "This museum's position is that the Nevada Test Site was a battleground of the Cold War and it helped to end it, and that's a significant position to take" (Friess 2005). Such a stance reduces the downwinders' status to that of merely unintended collateral damage on the front lines of the Cold War. ATM supporter Ernie Williams, former NTS controller, best summarized this attitude regarding the museum's mission: "Our museum is here to show what we did in defense of our country," said Williams (Friess 2005).

The Atomic Testing Museum demonstrates how local heritage preservation efforts can gain political support—if organizations and individuals focus solely on the government-sanctioned national narrative, victory in the Cold War. The museum attracted state, federal, and private corporate monies along with Smithsonian Institution affiliation. While heralding American Cold War victory gained the museum millions, it has also earned itself a vocal array of detractors. Downwinders have charged the museum with revising history. Even more moderate critics question the unbalanced story presented by the museum staff. Perhaps over time, the ATM will become more inclusive and accepting of the multiple narratives that make up the complete history of the Nevada Test Site.

FERNALD'S MULTI-USE EDUCATION FACILITY: TO BUILD OR NOT TO BUILD?

The Fernald uranium refinery, located twenty-five miles west of downtown Cincinnati, displaced 1,050 acres of farmland and houses (Wilson 1956). Beginning in the spring of 1951, the U.S. Army Corps of Engineers and AEC representatives informed fifty residents living and farming in this corner of southwestern Ohio that the federal government intended to acquire their lands for a new "atomic plant" (Wilson 1956). Fernald's proximity to a large, nearby labor market in Cincinnati, access to housing and transportation, and its central location between ore delivery ports and other production facilities in the arsenal made it an attractive location. Following the

Commission's official announcement, condemnation and eviction proceedings began around Fernald in 1951.

Fernald's selection reflected the changes taking place within the AEC. By 1951, the Commission determined that the previously dispersed uranium processing operations needed to be centralized in order to increase efficiency within the complex. The Department of Defense warned the Commission at the same time that only specific regions of the nation fell within the DOD's acceptable "First Defense Zone" (History Associates, Inc. 1987; Wilson 1956). This zone included most southern states along with portions of Kentucky, West Virginia, Indiana, and Missouri. This preferred defensive zone, according to 1950s estimates by the Defense Department, was likely out of range of Soviet bombers. Working within these parameters, Commission and contractor personnel identified Fernald as the best location for a uranium refinery.

Plant construction at Fernald began in 1951 and ended in 1954. When finished, the refinery complex consisted of ten production plants, twenty administrative, service and laboratory buildings, waste storage areas, and a buffer zone (Fluor Public Affairs Department 2001). AEC officials decided within a year of ending construction that Fernald operations needed to increase production and efficiency in order to meet rising demands from within the nuclear arsenal. The site achieved peak productivity in 1960 when it made 10,000 metric tons of uranium product (Golightly 1993). However, demand slowed over time as surplus feed materials (the government's domesticated term for uranium metal products) accumulated within the weapons complex. A series of production reactor closings, a large stockpile of plutonium devices, and alternative facilities capable of converting this reserve into new configurations raised questions concerning the need to continue operations at Fernald. Fernald faced declining orders from the government, workforce shrinkage, and equipment deterioration by the mid-1970s. When President Ronald Reagan called for a significant resurgence in the production of weapons-grade nuclear materials in the 1980s, the years of neglect and new strains severely strained Fernald's three-decade-old plants.

Beginning in the mid-1980s, nearby residents became aware of the health and environmental dangers posed by Fernald. A series of publicized incidents in late 1984 and 1985 added to the troubles of the Fernald site. These incidents contributed to the creation of the locally organized Fernald Residents for Environmental Safety & Health (FRESH). FRESH members began to pressure the DOE to end operations at the site. Public outrage, local media coverage, and state political action led the DOE to suspend operations in 1989 and to close the site permanently two years later. In December 1992, DOE leaders announced that it had granted the Fernald Environmental Restoration Management Corporation (FERMCO) the remediation contract for the site.

FERMCO was a subsidiary of Fluor-Daniel, Inc., part of the Fluor Corporation (Fernald Cleanup 1996; Gallagher 1996). FERMCO would operate as the primary cleanup contractor at Fernald over the entire course of site remediation, although the company changed names twice before settling on Fluor-Fernald in 1999.

Historic preservation did not emerge as a substantive issue until 1997. Between 1992 and 1997, FERMCO began dismantling 200 structures on the site. The DOE, the federal Environmental Protection Agency (EPA), and state regulators agreed on the construction of the On-Site Disposal Facility (OSDF) at Fernald in 1995. The OSDF design called for a width of 800 feet and a 3,700-foot length, with a 165-foot peak. The waste facility was built to contain 2.5 million cubic yards of contaminated and low-level radioactive materials. During July of the same year that DOE and EPA officials negotiated the OSDF agreement, a newly created and federally funded community advisory organization, Fernald Citizen's Task Force (FCTF), issued a report that included recommendations concerning the post-closure uses of Fernald. The FCTF specifically argued that the disposal cell should be incorporated into the local environment and minimally affect the area's vistas. In addition, the organization proposed barriers in the cell's design to be "as unobtrusive as possible, while providing clear markings to protect from intrusion" (FCTF 1995, 45). The only mention of heritage management in the report dealt with the re-interment of Native American remains (FCTF 1995, 46).

In 1997, the University of Cincinnati's Center for Environmental Communication Studies, in partnership with a handful of area residents and site workers, initiated efforts to preserve the history of Fernald. The Center developed the Fernald Living History Project (FLHP), an oral history program created to collect the remembrances of people who either worked or lived near Fernald from 1951 to the present (Barnes-Kloth et. al. 2001). The Project aimed to establish an archive of audio and videotaped interviews along with a compilation of photographs, letters, and memoirs from employees, residents, activists, and family members. FLHP leaders did not initially propose the construction of a museum-like facility, but they did strive to promote a broader understanding of the site's history and relevance to the public. Two years after its formation, the project, with the assistance of Fluor-Fernald, had collected over 100 videotaped interviews with retired workers and community members. The FLHP leadership incorporated the project in 1999, becoming the nonprofit Fernald Living History, Inc. (FLH).

Historical preservation languished at Fernald, despite the efforts by FLH to attract attention to its activities. The DOE, Fluor-Fernald, FRESH, and the Fernald Citizen's Advisory Board (FCAB), a reconstituted FCTF, devoted most of their time and money to remediation operations at the site. As remediation began to wind toward completion, FCAB slowly began to turn its attention to preserving aspects of site history. The citizen's advisory board sponsored a series of community "Future of Fernald Workshops" beginning in the spring of 1999. FCAB hoped that the workshops would encourage area residents to voice their ideas and concerns about the long-term post-closure development of Fernald. Based on input from the workshops, the FCAB adopted a vision statement in September 2000. FCAB proposed the construction of a "property that creates a federally-owned regional destination for educating this and future generations about the rich and varied history of Fernald" (FCAB 2002, 103). The advisory board's leadership strongly recommended that the DOE adopt this statement as a guide for future planning at the site.

Historical preservation and public education assumed a prominent position in FCAB's analysis of Fernald post-closure usage. In 2002, the advisory board released a report, *Telling the Story of Fernald*, that justified the need for a museum-like/multi-use educational facility. FCAB claimed that a multi-use center "would play an important integrating role in the future of the Fernald site, combining the wide variety of information, education, outreach, and long-term stewardship needs into one community-centered location" (FCAB 2002, 77). The board envisioned a facility consisting of an interpretive museum, an information/research hub, an educational/research center, and meeting rooms.

FCAB included in the report preliminary architectural drawings of the suggested center. The structure's design placed much of the building below grade (FCAB 2002). Tourists approaching the center by vehicle would drive down a ramp toward a parking lot that took them through layers of Fernald's history. Images and items representing the site's past were to be impressed into a retaining wall that guests would drive by on their way to the facility's front entrance. The retaining wall would continue into the structure, resurfacing in major interior public spaces, thus connecting the building and landscape together through Fernald's history.

The advisory board vehemently opposed the establishment of a smaller, single-use facility at Fernald. Members believed that such a reduced center would likely experience problems attracting visitors and retaining community interest. FCAB leaders argued in their report that "for a facility to be effective, it must become a sustaining force within the community, offering programs and activities that draw an ongoing regional audience to the site" (FCAB 2002, 77). The advisory board envisioned a center that both presented the technical history of Fernald production and remediation and preserved the diverse memories and artifacts of workers and residents. Such a facility would not only inform the public of the dangers still at Fernald, but would also advance knowledge about the harmful legacy of nuclear weapons production to future generations.

While the DOE acknowledged FCAB's concept, the agency made no effort to implement it. When DOE officials released the Comprehensive Stewardship Plan for Fernald in 2003, a document required for final closure, local residents and activists noticed that the document blatantly disregarded the FCAB proposal (USDOE 2003). Employing classic nukespeak, DOE officials noted that the provision of "a utilitarian type structure" on-site to house only necessary historical records would fulfill the agency's community obligations (USDOE 2003, 48). The Ohio Environmental Protection Agency and community organizations disagreed with the Energy Department on this matter (Klepal 2004; USDOE 2004a).

State and local criticisms directed at the DOE continued into 2004. The Ohio EPA severely attacked the DOE's July 2004 draft report on Fernald's legacy-management institutional controls (USDOE 2004a). State officials reiterated their support for a multi-use center while bashing the DOE's shortsightedness for relying solely on signs and fences to protect the disposal cell and public. The Ohio EPA noted in its formal comments that the DOE already operated facilities similar to that proposed at Fernald (USDOE 2004a, 6, 12). DOE personnel disputed this claim, stating that similar centers were regarded as public amenities and not requirements for legacy management by the agency (7, 12). The DOE promised, though, to provide several on-site trailers for "post-closure information including storage of photographs" (12). However, in December DOE personnel informed FCAB that the trailers might be shipped to the department's gaseous diffusion plant at Portsmouth, Ohio, for reuse (Depoe 2004b; Innis 2004).

Throughout 2004, the Ohio EPA and several local organizations worked to alter the DOE's stance regarding a multi-use education facility (MUEF) at Fernald. Fernald Living History, Inc. became embroiled in the struggle to persuade the DOE to commit to the center concept, and also tried to persuade the agency to preserve the photographic archive of the site before contractor Fluor-Fernald destroyed the negatives and pictures (Depoe 2004a). By the fall, the Ohio EPA, FCAB and FLH had been joined by the Hamilton County Environmental Action Committee and Hamilton County Solid Waste Management District in urging DOE personnel to reconsider the multi-use facility proposal (Alutto 2004; Schmucker 2004).

Initially, these protests seemed to produce some results. Paul Golan, DOE Principal Deputy Assistant Secretary for Environmental Management, publicly disclosed in December, 2004 that the department planned to "convert existing modular office buildings and a large warehouse into a campus-like setting" for a post-closure information center (Golan 2004). Golan further added that the facility might house records and "displays that will provide information on the full spectrum of Fernald's history" (Golan 2004).

Golan's statements indicated an apparent shift in the DOE's attitudes toward the MUEF concept, a move which continued into 2005. The DOE incorporated the MUEF into the agency's final draft of the legacy management plan for Fernald in September 2005. Johnny W. Reising, director of the DOE's Fernald Closure Project, restated the department's commitment to the multi-use structure in letters he wrote thanking community members for reviewing the September draft. He asserted in January 2006 that the "DOE agrees with the recommendation that the MUEF will be more than a building. The MUEF is envisioned to be a vital link to the community to communicate and maintain institutional controls at the site. The MUEF will also serve to educate the Public and Stakeholders about the history of the Fernald site and the region" (Reising 2006).

Two months later, in yet another twist to the MUEF saga, Ohio State Senior Assistant Attorney General Timothy J. Kern informed the U.S. Department of Justice that the Ohio Attorney General's Office wished to "delete the multi-use educational facility" from the settlement conditions of a Natural Resources Damage suit the state had filed earlier against the Energy Department pertaining to Fernald (Kern 2006). In a demonstration of unity, FCAB, FLH, and FRESH drafted a joint letter on March 7, 2006, encouraging DOE and state officials to resolve the lawsuit. The leadership hoped to encourage "both sides to this suit to settle in the best interests of the site and the community" and praised the DOE's apparent attitude shift, notably "identifying public involvement and a multi-use center as a key component of a legally-enforceable plan" (Bierer 2006). Despite this latest correspondence, the future of the MUEF remains uncertain.

When the DOE officially closes Fernald, very little physical evidence will visually disclose the extensive industrial complex that operated on the site. Wildlife and wetlands will characterize an area once checkered with production buildings, waste pits, offices, and roads. At present, the only lasting indications of the locale's prior contributions to the nation's defense consist of the disposal cell and monitoring stations. The DOE's plans call for the minimal use of signage near the waste tomb and even less throughout the rest of the site. When Fernald closes, its physical history will largely be entombed beneath the redeemed landscape (O'Farrell, 2006). Unless a MUEF is developed on the site with programming in environmental education and historic preservation envisioned by local stakeholders, the memories of Fernald will likely be forgotten within a couple of generations.

THE ROCKY FLATS COLD WAR MUSEUM: PROMISES UNFULFILLED

Colorado Front Range residents awoke on March 23, 1951 to read in local papers that the U.S. Atomic Energy Commission had decided to construct a nuclear weapons manufacturing plant just west of downtown Denver on the Rocky Flats mesa (Nover 1951). Immediate reactions tended toward euphoria as Denver and Boulder politicians and residents realized the boost to the local economy that this major federal spending commitment represented for the region. AEC commissioners planned to build a plant that recovered and recycled scrap plutonium, machined plutonium and highly enriched uranium parts, and produced plutonium pits, part of the trigger mechanism in nuclear warheads. While speculation surrounded Rocky Flats after the site construction, the public remained largely unaware of its exact purposes and potential hazards.

This changed in 1957. A major plutonium fire broke out in the glove box production line housed in Building 771 on September 11 (Ackland 1999; Dow Chemical 1957; Rocky Mountain Peace and Justice Center 1992). Building 771 was a primary production plant for manufacturing the plutonium triggering devices. Although employees put the fire out, they resorted to using water which raised the very real danger of causing a criticality incident. Both major metropolitan Denver newspapers ran articles the next day about the fire. AEC spokesmen tried to reassure the public that nothing hazardous escaped from the building. Local reporters failed to press the issue, and three weeks later public attention was diverted by the successful Soviet launch of Sputnik 1.

The 1957 fire became part of a pattern at Rocky Flats. Plutonium fires evolved into routine events at the site. Between 1954 and 1971, approximately 615 fires occurred at the site, including 384 in the plutonium production facilities (Rocky Mountain Peace and Justice Center 1992). The worst fire took place on Mother's Day, May 11, 1969, a blaze that surpassed the 1957 incident. Rags contaminated with plutonium flecks caught fire within a glove box on the north foundry line in Building 776/777, a production facility built in 1957. The fire sped quickly through the line, breaking the windows and glove box portholes. When the Rocky Flats fire department arrived, dense smoke, fireballs, and glowing droplets of lead greeted firefighters as they entered the production area. Gray and black smoke poured from exhaust vents on the building's roof by late afternoon (USAEC 1969). Drivers reported seeing the smoke plume from the Denver-Boulder Turnpike, approximately ten miles away from the site (USAEC 1969). An unknown amount of radioactive particulates escaped from the plant through the roof vents.

The 1969 fire made Denver residents more aware of the dangers posed by Rocky Flats and its operations. AEC personnel tried to allay public concerns, and reassured Front Range residents that little contamination actually escaped from the building despite the smoke plume (Huber 1969; Lindberg 1969). The Commission eventually admitted that the 1969 fire cost $45 million, plus an additional $22 million for the value of the plutonium consumed during the incident (Rocky Mountain Peace and Justice Center 1992).

The 1969 fire and the mounting evidence that production operations at Rocky Flats generated environmental and human health risks made the public reconsider the benefits of the site. Demonstrations started to occur outside the Rocky Flats security fence beginning in 1978. Protesters demanded the closure of the site and the disarmament of the nation's nuclear arsenal. The biggest demonstration took place on October 15, 1978, when nearly 20,000 people tried to symbolically encircle Rocky Flats by linking arms. Despite these incidents of public disgruntlement, production continued at Rocky Flats until 1989. FBI and federal EPA investigators raided Rocky Flats on June 6, 1989, based on reports of significant risks to worker and public safety at the site (Little 2005).

The 1989 raid resulted in the dismissal of Rockwell International as Rocky Flats' operations contractor and contributed to the permanent closure of the site in 1991. DOE officials made an attempt to reopen the site after the raid, but the overwhelming evidence of environmental, safety, and health violations torpedoed that effort. Rocky Flats' closure generated debate about the future of one of the most contaminated industrial landscapes in the country (Ackland 1999).

Following the suspension of operations in 1992, numerous local stakeholders provided competing visions of the site's future. The Rocky Flats Local Impact Initiative group, sponsored by the DOE, suggested the retention and conversion of non-radioactive buildings for commercial uses. Another organization, the Future Site Working Group, advocated limited public access to Rocky Flats after cleanup ended. Private real estate developers hoped to exploit the ten-square-mile plant site's scenic vistas of the Front Range. All these competing interest groups opposed the possibility of building an on-site disposal cell at Rocky Flats. DOE officials and personnel of EG&G, the contractor that replaced Rockwell, had suggested the waste cell idea in 1995 (Roberts 1995; Scanlon 1995). DOE officials remained committed to this idea until 1997, when adamant local dissent led by the federally funded Rocky Flats Citizens Advisory Board (RFCAB) forced Energy Department leaders to back off from on-site waste storage.

While government officials, residents and local activists debated the future uses of Rocky Flats, remediation efforts proceeded slowly at the site. By 1997, only a handful of minor production buildings had been demolished. Cleanup operations slowed even more when demolition moved into the primary plutonium production plant areas in 1998. The high levels of contamination found in this area forced the DOE and its cleanup contractor, Kaiser-Hill Company, to adopt a more cautious approach when remediating this zone. The slower cleanup pace, although frustrating for government officials still hoping to achieve their 2007 deadline, gave residents, activists and politicians an opportunity to more fully consider Rocky Flats' future.

In 1999, Representative Mark Udall (D-Colorado) and Senator Wayne Allard (R-Colorado) sparked a discussion about Rocky Flats' legacy, including the issue of preserving the site's history. They jointly introduced legislation that sponsored the creation of a National Wildlife Refuge at Rocky Flats once the Energy Department finished its cleanup operation (Udall's Off to a Good Start 1999). Embedded within the proposed bill was language that authorized the establishment of a "Rocky Flats Museum, to be located in Arvada, a western suburb of Denver" (U.S. Congress 2000).

Most local politicians, newspapers, and residents supported the conversion of Rocky Flats into open space. The Rocky Flats museum proposal touched off a firestorm, particularly from Arvada. Many Front Range politicos regarded Arvada's dissent as an indication that the city aspired to annex Rocky Flats for future development purposes (Gerhardt 2000; Steers 2000). By August, 2000, Arvada's mayor gave up the opposition and agreed to back the Udall-Allard legislation.

The Rocky Flats History Project had been founded in the meantime. The Project included representatives from the Arvada Center for the Arts and Humanities, the City of Arvada, Kaiser-Hill, and the Rocky Flats Homesteaders, an organization for retired site workers, as well as academics and community activists (Rocky Flats History Project 2000). The History Project intended to use the language in the Udall-Allard legislation to establish and fund a Cold War museum. The group identified and acquired artifacts from Rocky Flats' past. Members also planned to conduct an oral history project along with videotaping the critical issues surrounding the site. The History Project's long-term goal was constructing a museum to "commemorate and interpret the historical, scientific, and environmental legacy of Rocky Flats" (Rocky Flats History Project 2000). The project began organizing committees to consider collections/content, membership, by-laws, and nominations in 2001. The group's inventory of artifacts grew eclectically. The collection consisted of several training tools/instruments, including a demonstration glove box, mugs, magazines, buttons, and a variety of production-related materials.

Accelerated building demolition and public sale of non-radioactive items became growing concerns for the History Project (Romano 1999). In July 2001, the collection/content committee proposed that several acres within the Rocky Flats plant site be set aside for a museum. The committee suggested that the History Project locate an off-site storage facility with climate control to serve as an interim location for the expanding artifact collection. Organization members approved the incorporation of the Project and secured nonprofit status in the summer of 2001. The History Project leadership renamed the organization the Rocky Flats Cold War Museum (RFCWM).

The newly created museum board of directors requested $150,000 in seed money from Kaiser-Hill, payable over two years. The board planned to use the funds to "refine the goals and objectives of the project and implement the initial vision for the new museum" (Lyons 2001). The board leadership envisioned a facility that would serve multiple functions for the community. The RFCWM would "preserve representative examples of the buildings as artifacts" and initiate oral/video history projects to document the stories of Rocky Flats (Lyons 2001). Furthermore, the board argued that the museum could act as an archival repository and a key part of the long-term stewardship program for Rocky Flats. Kaiser-Hill leaders approved the funding request.

The efforts made by the RFCWM Board to propose a museum were matched by progress in designating Rocky Flats a wildlife refuge. Throughout 2001, Udall, Allard, and other Colorado politicians stumped for the passage of the Rocky Flats National Wildlife Refuge Act. While the legislation's sponsors anticipated a hard road to achieve passage, the act became fast-tracked by December 2001. The Colorado delegation attached the legislation onto a Defense Department spending bill. Congress approved the Udall-Allard act by mid-December, securing Wildlife Refuge status for Rocky Flats once remediation concluded at the site.

Wildlife refuge approval empowered the RFCWM Board to push ahead with historic preservation efforts. The organization hired a half-time executive director in June 2002. The Board also contracted with Informal Learning Experience, Inc., in 2002 to study the feasibility of building a museum at Rocky Flats. RFCWM released the study's findings in August, 2003. Informal Learning's analysts recommended locating the museum on or near the plant site, preferably in the existing visitor's center (West et al. 2003). The study outlined the need for the facility to comment on and interpret Rocky Flats' history due to the diversity of people, events, and issues connected to the site. The report projected that capital costs for museum construction ranged between $10 and 13 million and that the center might attract 110,000 visitors a year (West et al. 2003).

Museum proponents secured the support of other community organizations such as the Rocky Flats Citizens Advisory Board (RFCAB 2001). Both Udall and Allard approved the study's suggestions. Udall noted that Rocky Flats had contributed to the economic and community development of the Front Range region, stating that "this museum will be an important facility to commemorate that legacy while also promoting understanding, learning, and remembering" (Rocky Flats Cold War Museum 2003).

The excitement and enthusiasm generated by the report proved to be premature. DOE officials needed to submit a report to Congress by 2004 assessing the potential costs of the proposed Rocky Flats museum. The Udall-Allard bill required this submission, and DOE leaders wanted to respond to the RFCWM study. The DOE's Office of Legacy Management (DOE-LM) hired an independent museum consultant to analyze the situation at Rocky Flats. The consultant's conclusions served as the basis of the DOE's draft report. This draft study spawned angry reactions when DOE released it in November, 2004 for public comment.

DOE leaders acknowledged in the report that a museum was possible and could serve as an important tool in promoting public awareness of Rocky Flats and its hazards. However, the Energy Department study stated "it's DOE's opinion that a Rocky Flats Museum should only be established if it can be viable without federal funding" (USDOE 2004b, Executive Summary). Naturally, museum proponents reacted harshly to the Energy Department draft. RFCWM Board president Bryan Taylor and executive director Steven Davis quickly pointed out in a written statement that "numerous instances of successful federal and private funding partnerships for developing nuclear-historical facilities existed across the country" (Taylor 2004). The RFCAB leadership directed a highly critical letter to the Energy Department's Office of Legacy Management, noting "with some curiosity DOE's insistence that it not fund the museum at Rocky Flats when such funding is provided at other sites" (DePoorter 2004). Likewise, the Rocky Flats Coalition of Local Governments attacked the DOE for "actively working to eliminate a feasible option [the museum] at this time" (Imbierowicz 2004).

Despite a Congressional mandate, local community support, and two separate feasibility surveys, the Rocky Flats Cold War Museum group encountered the same DOE resistance as did MUEF supporters at Fernald. The Udall-Allard act provided formal endorsement for the *idea* of a museum, required a museum feasibility study, and seemed to short-circuit the usual bureaucratic bottlenecks. The Kaiser-Hill grant provided monies for a survey study and a part-time executive director, and placed the RFCWM on track toward task completion. However, the DOE's draft report dealt the RFCWM a body blow. When asked to comment further on the museum board's reaction to the DOE draft, Steven Davis commented, "During the past decade, it's become increasingly clear that DOE doesn't want to be in the museum business" (McGuire 2004, C-01).

The RFCWM currently exists only on paper and in cyberspace, although that may be changing. The organization has produced a series of eight display boards that offer a summary of the site's history along with an outline of the museum proposal. Along with the placards, RFCWM created a brochure further detailing the rationales for the project. These materials were made available for public access via the Internet. The organization continues to collect artifacts and documents related to Rocky Flats' history. The *Denver Post* reported in December 2005, that due to fundraising difficulties, the RFCWM Board might abandon the effort to purchase and renovate a building near Rocky Flats that the organization intended to use for the museum (McGuire 2005). By this time, it had become clear that when the U.S. Fish and Wildlife Service assumed management of the site after cleanup operations ended, there would be no on-site museum or information center for visitors (USFWS 2005).

The future brightened for the RFCWM movement in March, 2006. The organization's board announced on March 27 that the museum "received a donation of 1.4 acres of land near the former nuclear weapons production plant site" (Rocky Flats Cold War Museum 2006). The property was made available to the board on the condition that the museum be planned, financed, designed and ready for groundbreaking by January 1, 2008 (Rocky Flats Cold War Museum 2006). The donation came from board member Charles Church McKay, a local developer and nephew of Marcus Church, a prominent local figure. The Churches were one of the homesteading families on the Rocky Flats mesa evicted by the government in 1951.

When and if completed, the Rocky Flats Cold War Museum will likely owe its existence not to federal government subsidies, but to the determination and generosity of local activists, community leaders, and private individuals. The DOE's Office of Legacy Management consistently refuses to accept any obligation for heritage preservation and interpretation as it administers the closure of several former weapons complex sites. When the DOE passes authority for Rocky Flats over to the Fish and Wildlife Service, no physical plant structure will remain; the landscape and historical memory will have been sanitized by the government.

ALTERNATIVE POSSIBILITIES: ROCKY FLATS VIRTUAL HISTORY

When the efforts bogged down to build a Rocky Flats Cold War Museum, one of the individuals involved in the effort departed to construct a virtual museum of the site. Len Ackland, former editor of the *Bulletin of Atomic Scientists* and Associate Professor of Journalism at the University of Colorado, along with other interested parties, created an "on-line exhibit exploring a potentially catastrophic fire" at Rocky Flats (RFVM 2006a). Additionally, the supporters of this virtual exhibit hoped to "educate visitors by telling the extraordinary stories of the people, politics, science and secrecy of this former bomb factory" (RFVM 2006b). The web site (available at www.colorado.edu/journalism/cej/exhibit) focuses primarily on the 1969 Mother's Day fire discussed above, and also presents a "Nuclear-Age Timeline" spanning from 1942 to 2002 that includes significant Cold War and site-specific events.

The virtual exhibit provides a multimedia, hypertextual experience for visitors. Not only does the written content provide a gripping narrative of the fire, its nearly disastrous escape from the main plutonium pit production plant, and the immediate consequences of the event, but the website also includes photographs and audio clips of interviews of former plant workers. The site is interactive, with buttons placed throughout that allow viewers to read source documents pertaining to the fire and to hear the memories of those who encountered the fire personally. Significantly, these sources refute statements made at the time by AEC officials that radioactive particulates did not escape from the building during the fire and discuss the hazards firefighters encountered while putting the fire out. The virtual tour also explains to visitors how the glove box production lines operated at Rocky Flats, along with the pyrophilic characteristics of plutonium.

The key segment of the virtual tour concerns the potential criticality incident that could have happened during the fire. On a page entitled "Blue Flash of Death," which includes audio clips of workers' interviews along with images from official after-the-fact reports, visitors learn that conditions occurred during the fire that could have a triggered a chain reaction (RFVM 2006c). Such a criticality incident would have killed those immediately in the vicinity and seriously contaminated those outside of the production room area. The museum then chronicles the series of fortunate breaks that took place which likely prevented the fire from developing into a catastrophe for Front Range residents. While firefighters managed the fire within the building in the production area, others sprayed thousands of gallons of water onto the roof. This action kept the roof from collapsing and stopped the fire from escaping.

The next several pages in the virtual tour focus on the immediate consequences of the fire and the long-term impact of the event on local politics. The fire and official efforts to explain how this incident happened led to greater scrutiny of the conditions at the plant and more public awareness of the dangers Rocky Flats represented for residents. The museum claims that a "can-do" mentality and a work culture that accepted plutonium fires as inevitable contributed to the conditions that led up to the 1969 fire (RFVM 2006d). The tour then shifts focus to consider the broader implications of nuclear bomb production, noting that, despite the fire, production resumed at Rocky Flats within a few months.

The final page of the virtual museum refocuses on the local reaction to the 1969 fire. The tour states that if local residents "had known the full truth about the risks to which they and their families were being subjected, they would have been outraged" (RFVM 2006e). The museum also makes the dramatic (and perhaps questionable) claim that the 1969 fire "could have caused a Chernobyl-scale disaster" (RFVM 2006f).

The virtual Rocky Flats exhibit offers an alternative version of the 1969 fire. The often-sanitized accounts of the incident used by the federal government downplay and ignore the nearly disastrous outcome of the fire. The web pages of the museum tour provide provocative content through narrative context, audio clips from interviews, and video segments from later television investigations. The titles of the individual web pages themselves are designed to evoke strong audience reactions. Page One, for example, is entitled "The Day We Almost Lost Denver," while the header on Page Five reads "Blue Flash of Death." These titles challenge the traditional framework of the fire story, providing the audience with a different perspective. When combined with the images used throughout the tour, the virtual tour works to undermine the officially sanctioned version of the fire. Ironically, many of the visual aids on the pages actually come from government sources.

The Rocky Flats Virtual Exhibit acts as a guerrilla museum, a counter-statement much like the counter-monuments erected in Germany in recent years to protest the government's version of World War II (Young 1992). The site presents an alternative perspective not only of the nuclear weapons complex, but also of the ideological history of the Cold War. The museum displays the darker aspects of the nation's forty-five year standoff with the Soviet Union, aspects ignored by the American Testing Museum in Nevada. There is no Cold War triumphalism in the Rocky Flats Virtual Museum, no Ground Zero Theater—only the story of a nearly catastrophic loss of life due ultimately to human hubris. Only outside the pale of federal funding could a museum, even an online version like the Rocky Flats Virtual Exhibit, be possible.

CONCLUSION: REMEMBERING IN TRIUMPH, FORGETTING IN SILENCE

The federal government has spent hundreds of billions of dollars constructing and operating the nuclear weapons complex. These activities have taken place largely out of public view. Only in recent years have the extent of these operations and the complexity of the nuclear arsenal begun to surface. Now, localized efforts to preserve site histories are inadequately supported, if not suppressed, by the federal government. Those stories that are being preserved by the DOE reinforce the officially sanctioned story of an America triumphant in the Cold War.

What the DOE has left behind is a mosaic of atomic spaces (Hales 1997), a complex in which some sites present celebrated—and manipulated—accounts of Cold War history while other sites present nothing at all (Molella 2003). As a result, much of the American public remains uninformed about the health risks and environmental dangers that remain from the legacy of nuclear weapons production. Our collective memory of the Cold War and the continuing impact of nuclear weapons on our lives remains incomplete and distorted. Perhaps alternative projects such as the Rocky Flats Virtual Museum can provide a much-needed critical history to offset the combination of Cold War heritage preservation at places like the Atomic Testing Museum, and the attempted erasure of the historical record at places such as Fernald and Rocky Flats.

NOTES

1. Jason Krupar would like to thank his wife, Carmen, for her continual support of his research and writing. Stephen Depoe would like to thank his co-author for his expertise and patience.
2. In fact, DOE and contractors completed much of the remediation of Rocky Flats by 2006, and declared "closure" of the site. The anticipated "closure" date at Fernald is late 2006 or early 2007.

BIBLIOGRAPHY

Ackland, Len. (1999). *Making a real killing: Rocky Flats and the nuclear west*. Albuquerque: University of New Mexico Press.

Aluotto, Jeffrey W. 2004, November 18. Letter to William J. Taylor, DOE-Fernald.

Barnes-Kloth, Rhonda, Stephen Depoe, Jennifer Hamilton, and Amy Lombardo. 2001. Memories of Fernald: Defining a "sense of place" through personal narrative. Pp. 303–317 in *Proceedings of the fifth biennial conference on communication and the environment*, ed. Brant Short and Dayle Hardy-Short. Flagstaff: Northern Arizona University.

Beichman, Arnold, ed. 2000. *CNN's Cold War documentary: Issues and controversy*. Stanford, Calif.: Hoover Institution Press.

Bierer, Jim, Lisa Crawford, Joyce Bentle, and Susan Verkamp. 2006, March 7. Letter to Timothy J. Kern, Ohio Environmental Protection Agency, and Daniel Dertke, U.S. Department of Justice.

Blair, Carole, Marsha Jeppeson, and Enrico Pucci, Jr. 1991. Public memorializing in modernity: The Vietnam Veterans Memorial as prototype. *Quarterly Journal of Speech* 77: 263–288.

Bodnar, John. 1992. *Remaking America: Public memory, commemoration, and patriotism in the twentieth century*. Princeton, NJ: Princeton University Press.

Boyle, Peter. 2000. The Cold War revisited. *Journal of Contemporary History* 35: 479–489.

Brooks, Linton. 2005, February 19. Preserving the legacy: Dedication of the Atomic Testing Museum. Available from: www.nnsa.doe.gov/docs/speeches/2005/speech_Dedicate_Atomic_Testing_Museum.pdf (1 June 2006).

Browne, Stephen.1993. Reading public memory in Daniel Webster's *Plymouth Rock* oration. *Western Journal of Communication* 57: 464–477.

———. 1995. Reading, rhetoric, and the texture of public memory. *Quarterly Journal of Speech* 81: 237–250.

———. 1999. Remembering Crispus Attucks: Race, rhetoric, and the politics of commemoration. *Quarterly Journal of Speech* 85: 169–187.

Bruner, M. Lane. 2000. Strategies of remembrance in pre-unification West Germany. *Quarterly Journal of Speech* 86: 86–107.

Carpenter, Ronald. 1995. *History as rhetoric: Style, narrative, and persuasion*. Columbia, S.C.: University of South Carolina Press.

Clark, E. Culpepper, and Raymond McKerrow. 1998. The rhetorical construction of history. Pp. 33–46 in *Doing rhetorical history: Cases and concepts*, ed. Kathleen Turner. Tuscaloosa: University of Alabama Press.

Cox, J. Robert. 1990. Memory, critical theory, and the argument from history. *Argumentation & Advocacy* 27: 1–13.

Cox, Michael, ed. 1998. *Re-thinking the soviet collapse: Sovietology, the death of Communism and the new Russia*. London: Pinter Publishing.

Dean, Gordon. 1951, May 16. Letter to Senator Robert Taft. U.S. Department of Energy Archives, 326 U.S. Atomic Energy Commission, Gordon Dean Reading File, Job 1110, Box 0003, Folder 18.

Depoe, Stephen.1994. *Arthur Schlesinger, Jr., and the ideological history of American liberalism*. Tuscaloosa: University of Alabama Press.

———. 2004a, June 30. Letter to William J. Taylor, U.S. Department of Energy-Ohio Field Office.

———. 2004b, December 3. Letter to Lisa Crawford and Pam Dunn.

DePoorter, Gerald L. 2004, December 2. Letter to Ray Plieness.

Desert Research Institute. 2003. DRI Officially Opens the Frank H. Rogers Science and Technology Building today. Available at news.dri.edu/nr2003/oct_building.htm (15 March 2005).

Dickinson, Gary. 1997. Memories for sale: Nostalgia and the construction of identity in old Pasadena. *Quarterly Journal of Speech* 83: 1–27.

Dickson, Mary. 2005, February 27. Dickson: Atomic Museum ignores human toll of nuclear testing. *Salt Lake Tribune*, A–9.

Dow Chemical Company. 1957, October 7. *Unclassified version of the report of investigation of serious incident in Building 71 on September 11, 1957*. Midland, Mich.: Dow Chemical Company.

Downwinders charge Atomic Testing Museum of revisionist history. 2005, February 23. *Common Dreams Progressive Newswire*. Available from http://www.commondreams.org/news2005/0223-04.htm (1 August 2006).

Ehrenhaus, Peter. 2001. Why we fought: Holocaust memory in Spielberg's *Saving Private Ryan. Critical Studies in Mass Communication* 18: 321–337.

Fernald Citizens Advisory Board (FCAB). 2002. *Telling the story of Fernald.* Fernald, Ohio: Fernald Citizens Advisory Board.

Fernald Citizens Task Force (FCTF). 1995. *Recommendations on remediation levels, waste disposition, priorities, and future use.* Fernald, Ohio: Fernald Citizens Task Force.

Fernald cleanup company fixes up its name also. 1996, September 14. *Cincinnati Enquirer*, B–1.

Fluor Public Affairs Department. 2001. *Fernald at 50: From weapons to wetlands.* Fernald, Ohio: Fluor Fernald.

Foss, Karen, and Stephen Littlejohn. 1986. *The day after*: Rhetorical vision in an ironic frame. *Critical Studies in Mass Communication* 3: 317–336.

Friess, Steve. 2005, February 25. Nevada tips its hat to its atomic history. *Christian Science Monitor*. Available at www.csmonitor.com/2005/0225/p03s01-ussc.html (1 June 2006).

Gaddis, John Lewis. 1997. *We now know: Rethinking Cold War history.* New York: Oxford University Press.

Gallagher, Mike. 1996, August 7. Fernald firm gets contract to clean up Hanford. *Cincinnati Enquirer*, A–10.

Gallagher, Victoria. 1995. Remembering together: Rhetorical integration and the case of the Martin Luther King, Jr. memorial. *Southern Communication Journal* 60: 109–119.

Gerhardt, Gary. 2000, March 6. Arvada eying Flats land? *Rocky Mountain News*, n.p.

Glass, Matthew. 1993. *Citizens against the MX: Public languages in the nuclear age.* Urbana: University of Illinois Press.

Golan, Paul. 2004, December 21. Letter to Bruce Schmucker.

Golightly, J. 1993. *Site history of the Fernald Environmental Management Project.* Rockville, Md.: History Associates, Inc.

Gronbeck, Bruce. 1998. The rhetorics of the past: History, argument, and collective memory. Pp. 47–60 in *Doing rhetorical history: Cases and concepts*, ed. Kathleen Turner. Tuscaloosa: University of Alabama Press.

Gusterson, Hugh. 2004. Nuclear tourism. *Journal for Cultural Research* 8: 23–31.

Hales, Peter. 1997. *Atomic spaces: Living on the Manhattan Project.* Champaign: University of Illinois Press.

Hasian, Marouf, Jr. 2001a. The advent of critical memory studies and the future of legal argumentation. *Argumentation & Advocacy* 38: 40–45.

———. 2001b. Nostalgic longings: Memories of the "Good War," and cinematic representations in *Saving Private Ryan. Critical Studies in Mass Communication* 18: 338–358.

———. 2004. Remembering and forgetting the "Final Solution": A rhetorical pilgrimage through the U.S. Holocaust Memorial Museum. *Critical Studies in Mass Communication* 21: 64–92.

———. 2005. Authenticity, public memories, and the problematics of post-holocaust remembrances: A rhetorical analysis of *The Wilkominski Affair. Quarterly Journal of Speech* 91: 231–263.

Hasian, Marouf, Jr., and A. Cheree Carlson. 2000. Revisionism and collective memory: The struggle for meaning in the *Amistad* affair. *Communication Monographs* 67: 42–62.

Hasian, Marouf, Jr., and Robert Frank. 1999. Rhetoric, history, and collective memory: Decoding the Goldhagen debates. *Western Journal of Communication* 63: 95–114.

History Associates, Inc. 1987. *History of the production complex: The methods of site selection*. Rockville, Md.: History Associates, Inc.

Hogan, Michael, ed. 1992. *The end of the Cold War: Its meaning and implications*. New York: Cambridge University Press.

Hubbard, Bryan, and Marouf Hasian, Jr.1998. Atomic memories of the Enola Gay: Strategies of remembrance at the National Air and Space Museum. *Rhetoric and Public Affairs* 1: 363–385.

Huber, Bob. 1969, June 8. No contamination reported. *Denver Post*, n.p.

Hunter, Allen, ed. 1998. *Re-thinking the Cold War*. Philadelphia: Temple University Press.

Imbierowicz, Karen. 2004, December 6. Letter to Mike Owen, DOE.

Innis, James. 2004, December 3. Letter to Steve Depoe.

Jorgensen-Earp, Cheryl, and Lori Lanzilotti. 1998. Public memory and private grief: The construction of shrines at the sites of public tragedy. *Quarterly Journal of Speech* 84: 150–170.

Kane, Thomas. 1988. Rhetorical histories and arms negotiations. *Journal of the American Forensic Association* 24: 143–154.

Katriel, Tamar. 1993. "Our future is where our past is": Studying heritage museums as ideological and performative arenas. *Communication Monographs* 60: 69–75.

——. 1994. Sites of memory: Discourses of the past in Israeli pioneering settlement museums. *Quarterly Journal of Speech* 80: 1–20.

Katsiaficas, George, ed. 2001. *After the fall: 1989 and the future of freedom*. New York: Routledge.

Kauffman, Charles. 1989. Names and weapons. *Communication Monographs* 56: 273–285.

Kern, Timothy J. 2006, March 2. Letter to Daniel J. Dertke, U.S. Department of Justice.

Kimberly, Robert. 2005, 4 March. This is not a test: The Atomic Testing Museum is ground zero for a discussion of America's nuclear industry. *Las Vegas City Life*. Available at www.lasvegascitylife.com/articles/2005/03/03/art_page/art.txt (1 June 2006).

Kinsella, William J. 2001. Nuclear boundaries: Material and discursive containment at the Hanford nuclear reservation. *Science as Culture* 10: 163–194.

——. 2005. One hundred years of nuclear discourse: Four master tropes and their implications for environmental communication. Pp. 49–72 in *The environmental communication yearbook, vol. 2*, ed. Susan L. Senecah. Mahwah, N. J.: Erlbaum.

Klepal, Dan. 2004, July 5. Park's price tag disputed. *Cincinnati Enquirer*, B–1.

Kotkin, Stephen. 2001. *Armageddon averted: The Soviet collapse, 1970–2000*. New York: Oxford University Press.

Kuletz, Valerie. 1998. *The tainted desert: Environmental and social ruin in the American West*. London: Routledge.

Lake, Richard. 2005, February 26. Atomic Testing Museum: Downwinders feel left out. *Las Vegas Review-Journal*, 2–B.

Lindberg, Gene. 1969, August 23. Flats radiation listed at low level. *Denver Post*, n.p.

Little, Amanda. 2005, January 21. The Rocky Flats horror picture show. *Grist Magazine*. Available at www.grist.org/news/muck/2005/01/21/little-rockyflats/ (28 August 2006).

Lowenthal, David. 1998. *The heritage crusade and the spoils of history*. Cambridge: Cambridge University Press.

Lyons, Carol. 2001, July 16. Rocky Flats Cold War Museum: A proposal to Kaiser-Hill Company. Letter to Nancy Tuor.

Mandziuk, Roseann. 2003. Commemorating Sojourner Truth: Negotiating the politics of race and gender in the spaces of public memory. *Western Journal of Communication* 67: 271–291.

Manning, Mary. 2005, February 21. Protestors make stand at atomic museum opening. *Las Vegas Sun*. Available at www.lasvegassun.com/sunbin/stories/sun/2005/feb/21/518326384.html?ATOMIC%20MUSEUM%20daterange:2453395 2453760 (1 June 2006).

McGee, Michael C. 1977. The fall of Wellington: A case study of the relationship between theory, practice, and rhetoric in history. *Quarterly Journal of Speech* 63: 28–42.

McGuire, Kim. 2004, November 14. Flats museum is on its own for funds, feds say. *Denver Post*, C–01.

———. 005, December 29. Rocky Flats artifacts lack funds for home. *Denver Post*, B–01.

Molella, Arthur. 2003. Exhibiting atomic culture: The view from Oak Ridge. *History and Technology*, 19 (3): 211–226.

Myers, Dennis. 2005, March 3. Nuclear testing: Downplayed downwinders? *Reno News and Review*. Available at www.newsreview.com/reno/Content?oid=24874 (1 June 2006).

Nevada Test Site Historical Foundation. 2005a, March 15. Lockheed Martin donates $100,000 to NTSHF. Available at www.ntshf.org/lockheed%20news.htm (1 June 2006).

———. 2005b, March 15. Atomic Testing Museum opens in Las Vegas. Available at www.ntshf.org/atmopen.htm (1 June 2006).

———. 2005c, March 15. Atomic Testing Museum: Local Smithsonian affiliate contracts for installation of multi-million-dollar exhibits. Available at www.ntshf.org/atmnewsrel.htm (1 June 2006).

———. 2006, April 25. Atomic Testing Museum facts. Available at www.ntshf.org/atmfacts.htm (1 June 2006).

Nichols, Thomas. 2002. *Winning the world: Lessons for America's future from the Cold War*. Westport, Conn.: Praeger.

Nover, Barnett. 1951, March 23. U.S. to build $45 million A-plant near Denver. *Denver Post*, n.p.

O'Farrell, Peggy. 2006, July 23. Fernald is clean. *Cincinnati Enquirer*, C–1, C–6.

Parry-Giles, Shawn, and Trevor Parry-Giles. 2000. Collective memory, political nostalgia, and the rhetorical presidency: Bill Clinton's commemoration of the march on Washington, August 28, 1998. *Quarterly Journal of Speech,* 86: 417–437.

Phillips, Kendall, ed. 2004. *Framing public memory*. Tuscaloosa: University of Alabama Press.

Prosise, Theodore.1998. The collective memory of the atomic bombings misrecognized as objective history: The case of public opposition to the National Air and Space Museum's atomic bomb exhibit. *Western Journal of Communication* 62: 316–347.

Ray Wilson, James. 1956. *The Fernald project of the Atomic Energy Commission: An agricultural-industrial transformation*. Columbus: Ohio State University. M.A. thesis.

Reising, Johnny W. 2006, January 26. Letter to Steve Depoe, Susan J. Verkamp, Lisa Crawford, and James Bierer.

Roberts, Chris.1995, November 2. Flats plan: Raze, bury site. *Boulder Daily Camera*, n.p.

Robin, Corey. 2004. Remembrance of empires past: 9/11 and the end of the Cold War. Pp. 274–297 in *Cold War triumphalism: The misuse of history after the fall of Communism*, ed. Ellen Schrecker. New York: The New Press.

Rocky Flats Citizens Advisory Board (RFCAB). 2001, April 13. Meeting minutes.

Rocky Flats Cold War Museum. 2003, April 21. Rocky Flats museum one step closer. Press release.

——. 2006, March 27. Rocky Flats Cold War Museum secures land for museum site. Press release.

Rocky Flats History Project. 2000, June. A Cold War memorial: Update. Press release.

Rocky Flats Virtual Museum (RFVM). 2006a. About us. Available at www.colorado.edu/journalism/cej/exhibit/aboutus.html (29 June 2006).

——. 2006b. Mission. Available at www.colorado.edu/journalism/cej/exhibit/mission.html (29 June 2006).

——. 2006c. Blue Flash of Death. Available at www.colorado.edu/journalism/cej/exhibit/1969fire05-04.html (29 June 2006).

——. 2006d. The Fire Was Inevitable. Available at www.colorado.edu/journalism/cej/exhibit/1969fire09-04.html (29 June 2006).

——. 2006e. Citizens Awaken. Available at www.colorado.edu/journalism/cej/exhibit/1969fire11-04.html (29 June 2006).

——. 2006f. The Day We Almost Lost Denver. Available at www.colorado.edu/journalism/cej/exhibit/1969fire01-08.html (29 June 2006).

Rocky Mountain Peace and Justice Center. 1992. *Citizen's guide to Rocky Flats*. Boulder, Colo.: Rocky Mountain Peace and Justice Center.

Rogers, Keith. 2005, February 14. Museum a blast from the past: Exhibits document genesis, evolution of Nevada Test Site. *Las Vegas Review-Journal*, 1–B.

Romano, Michael. 1999, January 27. Bargain day coming up at Rocky Flats. *Rocky Mountain News*, n.p.

Rothstein, Edward. 2005, February 23. Museum Review: A place to consider apocalypse. *The New York Times*. Available at travel2.nytimes.com/2005/02/23/arts/design/23vega.html?ex=115654800&en=908638efc2488e8a&ei=5070 (29 June 2006).

Scanlon, Bill. 1995, January 27. Rocky Flats proposes burying waste on-site. *Rocky Mountain News*, n.p.

Schiappa, Edward. 1989. The rhetoric of nukespeak. *Communication Monographs* 56: 253–272.

Schmucker, Bruce. 2004, November 18. Letter to Christopher Jones, Ohio EPA, and William J. Taylor, DOE-Fernald.

Schrecker, Ellen. 2004. Introduction: Cold War triumphalism and the real Cold War. Pp. 1–24 in *Cold War triumphalism: The misuse of history after the fall of Communism*, ed. Ellen Schrecker. New York: The New Press.

Smith, Jeff. 1989. *Unthinking the unthinkable: Nuclear weapons and Western culture*. Bloomington: Indiana University Press.

Steers, Stuart. 2000, August 10. Forbidden Fruit. *Denver Westword*. Available at search.westword.com/Issues/2000-08-10/news/feature2.html (1 August 2006).

Suri, Jeremi. 2002. Explaining the end of the Cold War: A new historical consensus? *Journal of Cold War Studies* 4: 60–92.

Taylor, Bryan C. 1993a. Register of the repressed: Women's voice and body in the nuclear weapons organization. *Quarterly Journal of Speech* 79: 267–285.

———. 1993b. *Fat Man and Little Boy*: Cinematic representation of interests in the nuclear weapons organization. *Critical Studies in Mass Communication* 10: 367–394.

———. 1996. Make bomb, save world: Reflections on dialogic nuclear ethnography. *Journal of Contemporary Ethnography* 25: 120–143.

———. 1997a. Home zero: Images of home and field in nuclear-cultural studies. *Western Journal of Communication* 61: 209–234.

———. 1997b. Revis(it)ing nuclear history: Narrative conflict at the Bradbury Science Museum. *Studies in Cultures, Organizations, and Societies* 3: 119–145.

———. 1997c. Shooting downwind: Depicting the radiated body in epidemiology and documentary photography. Pp. 289–328 in *Transgressing discourses: Communication and the voice of other*, ed. Michael Huspek and Gary Radford. Albany: SUNY Press.

———. 1998a. Nuclear weapons and communication studies: A review essay. *Western Journal of Communication* 62: 300–315.

———. 1998b. The bodies of August: Photographic realism and controversy at the National Air and Space Museum. *Rhetoric and Public Affairs* 1: 331–361.

———. 2003a. "Our Bruised Arms Hung Up as Monuments": Nuclear iconography in post–Cold War culture. *Critical Studies in Media Communication* 20: 1–34.

———. 2003b. Nuclear waste and communication studies (Review of Thomas V. Peterson, *Linked arms: A rural community resists nuclear waste*). *Review of Communication* 3: 285–291.

Taylor, Bryan C. and Brian Freer. 2002. Containing the nuclear past: The politics of history and heritage at the Hanford plutonium works. *Journal of Organizational Change Management,* 15: 563–588.

Taylor, Bryan C. and Steven Davis. 2004, December 6. Letter to Mike Owen, DOE.

Taylor, Bryan C., William J. Kinsella, Stephen P. Depoe, and Maribeth S. Metzler. 2005. Nuclear legacies: Communication, controversy, and the U.S. nuclear weapons production complex. Pp. 363–409 in *Communication yearbook, vol. 29*, ed. Pamela Kalbfleisch. Mahwah, N.J.: Erlbaum.

Titus, Alice Costandina (Dina). 2001. *Bombs in the backyard: Atomic testing and American politics*. 2nd ed. Reno: University of Nevada Press.

Turner, Kathleen, ed. 1998. *Doing rhetorical history: Cases and concepts*. Tuscaloosa: University of Alabama Press.

Udall's off to a good start. 1999, June 14. Editorial. *Denver Post*, B–09.

U.S. Atomic Energy Commission (USAEC). 1969. *Report on investigation of fire building 776–777 Rocky Flats plant, volume I*. Washington, D.C.

United States Congress. 2000. *Rocky Flats National Wildlife Refuge Act of 2000*. Section 10. 106th Congress, 2nd Session. Washington, D.C.: USGPO.

U.S. Department of Energy (USDOE). 1997. *Linking legacies: Connecting the Cold War nuclear weapons production processes to their environmental consequences*. DOE-EM. Washington, D.C.: USGPO.

——. 2003. *Draft comprehensive stewardship plan: Fernald Closure Project, Fernald, Ohio*. DOE-EM. Washington, D.C.: USGPO.

——. 2004a. *Responses to U.S. and Ohio Environmental Protection Agency comments on the Draft Comprehensive Legacy Management and Institutional Controls Plan: Fernald Closure Project*. DOE-EM. Washington, D.C.: USGPO.

——. 2004b. *Draft report to Congress on Rocky Flats museum options*. DOE-Legacy Management. Washington, D.C.: USGPO.

U.S. Fish and Wildlife Service (USFWS). 2005. *Rocky Flats National Wildlife Refuge: Record of decision, Final Comprehensive Conservation Plan*. Washington, D.C.: USGPO.

Vanderbilt, Tom. 2002. *Survival city: Adventures among the ruins of atomic America*. New York: Princeton Architectural Press.

Wertsch, James V. 1987. Nuclear discourse. *Communication Research* 14: 131–138.

West, Robert M., Dan Martin, and Patrick Gallagher. 2003. *Rocky Flats Cold War Museum scoping study, final report.* Washington, D.C.: Informal Learning Experiences, Inc.

Young, James. 1992. The counter-monument: Memory against itself in Germany today. *Critical Inquiry,* 18 (2): 267–296.

Zelizer, Barbie. 1995. Reading the past against the grain: The shape of memory studies. *Critical Studies in Mass Communication*, 12: 214–235.

6

TRUTH Is Generated HERE

Knowledge Loss and the Production of Nuclear Confidence in the Post–Cold War Era

Laura A. McNamara

> Whenever a scientist has a very serious message to convey, he faces a problem of disbelief. How to be credible?
>
> —Mary Douglas, *Implicit Meanings*

During the past decade, scientists and engineers in the U.S. Department of Energy's (hereafter, DOE) nuclear weapon design laboratories have been engaged in an often-heated epistemological debate: Given current restrictions on the design and testing of nuclear explosives, what is the best way to maintain confidence in the United States' nuclear stockpile?

Throughout the Cold War, Los Alamos and its sister institutions, Lawrence Livermore and Sandia National Laboratories, produced confidence in the nuclear deterrent through an iterative cycle of designing, testing, refining and stockpiling nuclear explosives. That cycle, which structured work practices at the national laboratories for forty-seven years, was abruptly truncated in July of 1992, when Congress approved the Hatfield-Exon-Mitchell Amendment to the Energy and Water Appropriations Act. Within a few months, funding for the DOE's underground nuclear testing program evaporated, and the laboratories' core experimental program was quite literally left hanging, with massive experimental assemblies suspended in mid-completion over the dry desert floor of the Nevada Test Site (NTS).

In the wake of the test moratorium—which was renewed indefinitely in 1995—the DOE's nuclear weapons experts have been working feverishly to develop a new certification paradigm that will allow them to maintain confidence in the nuclear deterrent without full-scale testing. Known as Science Based Stockpile Stewardship (SBSS), the program provides the nuclear weapons laboratories several billion dollars per year to pursue a predictive capability grounded in basic physics, using a suite of advanced computational, experimental, and visualization tools.

Generous fiscal support notwithstanding, the transition from Cold War empiricism to the first principles approach embodied in SBSS has been neither smooth nor easy for the weapons laboratories. Each year, they are required to certify the continued reliability and performance of the seven nuclear weapon systems that make up the United States' enduring stockpile without conducting any nuclear tests. Compounding the difficulty of this task is the specter of knowledge loss: since 1992, weapon designers and engineers have been worrying that crucial skills and understandings, tacit ways of knowing nuclear weapons, are disappearing as they retire. Without an active program of nuclear weapons work, there are serious concerns that weapons knowledge is quite literally facing an imminent (and, from the perspective of most nuclear weapons experts, untimely) demise.

The knowledge loss issue was pinpointed in a provocative essay by Donald MacKenzie and Graham Spinardi, for whom the nuclear weapons laboratories' dilemma raised the prospect of cognitive decay in the nuclear weapons complex. In 1995, three years after the test moratorium, they suggested that a permanent ban could, in some limited sense, bring about the gradual disappearance of nuclear weapons:

> As designers themselves age, leave and die, the number who have first-hand experience of development through the point of full nuclear testing will steadily diminish, yet they will have to decide whether the inevitable changes in the arsenal matter. In such a situation, will explicit knowledge be enough? Will tacit knowledge and judgment survive adequately? For how long? (Mackenzie and Spinardi 1995, 92).

For the authors, the precarious state of post–Cold War nuclear expertise provides an opportunity to challenge the "traditional view" of science, in which knowledge is treated as "independent of context, impersonal, public and cumulative" (1995, 44). An alternative account of scientific knowledge posits that formal method may be less important than *tacit knowledge*: those ways of knowing that resist formal expression and must therefore be transmitted relationally and experientially. If tacit knowledge is indeed necessary for maintaining technological expertise, and the current moratoria on the design and testing of nuclear explosives make nuclear confidence dependent on explicit forms of weaponeering knowledge, argue the authors, then there is some sense in which nuclear weapons—or at least confidence in those weapons—may be "uninvented" (1995, 47).

Mackenzie and Spinardi are not the first to make this argument. Since the early 1950s, when a test ban was first suggested, its proponents have argued that a CTBT would slow the arms race by limiting its signatories' technical capability to develop new nuclear weapons. In complementary fashion, test ban opponents argued that testing was a necessary condition for maintaining the basis of technical expertise that underlies confidence in the nuclear deterrent (Birely 1987; Brown 1986). Nearly fifteen years after the moratorium, however, however, the United States maintains a substantial cache of nuclear explosives. Considering the impacts of the design and test moratoria, the exodus of experienced personnel, and the tremendous technical challenges presented by a rapidly aging stockpile, the United States' nuclear weapons programs are remarkably healthy. There have been no open challenges to the credibility of the nuclear deterrent, despite the fact that many of the most experienced designers and engineers have left the weapons programs. Moreover, deep cuts in the nuclear stockpile notwithstanding, the United States has not made significant progress toward developing a national security policy that does not rely on nuclear deterrence. Indeed, Congress is currently contemplating the development of a new nuclear system, the Reliable Replacement Warhead (RRW), which the weapons laboratories intend to certify without nuclear testing. The nuclear weapons laboratories, those quintessential relics of the Cold War, may be poised for something of a comeback.

In short, the terrain of weaponeering is simultaneously more stable and more contested than Mackenzie and Spinardi imagined. To explain this apparent paradox, it is necessary to broaden the scope of tacit knowledge, to understand the importance of community, identity and meaning in reinscribing a discursive regime of truth (Foucault 1981) in which nuclear weapons are a beneficial proscriptive technology, the ultimate arbiter in a world of conflict. Despite the fact that the mode of producing nuclear confidence has changed dramatically, perhaps irrevocably, with the end of nuclear testing, Los Alamos and its sister laboratories continue to play a key role in maintaining a regime of truth in which nuclear weapons—and thus the weapons community—remain critical for national security. Nuclear deterrence remains a potent motivator for the efforts of several thousand physicists, technicians, engineers, statisticians, metallurgists, chemists, computer scientists, and other researchers and technical experts who have spent the past decade working to sustain confidence in a stockpile of aging weapons that cannot be tested. In working to establish new "ways of knowing" nuclear weapons, weapons experts are redefining the very nature of nuclear confidence at a time when the role of nuclear weapons in national security is itself undergoing rapid change.

COMMUNICATION, ANTHROPOLOGY, AND WAYS OF KNOWING ABOUT NUCLEAR WEAPONS

Like their counterparts in communication, cultural anthropologists are interested in the study of "social things" (Lemert 2001) as a means of understanding how the actions of individuals reproduce and extend the social structures that they inherit at birth. This is a core research problem for all social scientists, methodological differences and disciplinary boundaries notwithstanding (Lukes 1982). Within the social sciences, anthropology is a quintessentially local discipline that has its roots in the long-term observation of small-scale, non-Western societies. The central organizing principle is firsthand participation and observation in an unfamiliar *culture*: the constellation of collectively held, experientially transmitted traits and behaviors that emerge as people develop ways to make sense of each other and the world around them. Cultural reproduction occurs as newcomers to a community gradually learn the geography, norms, practices, symbols, organization, hierarchy, language, frames, and tools that characterize the group. Over time, newcomers learn how to behave like a full member of the group, and reinscribe and extend what they have learned through their own action-in-the-world. To study culture, the anthropologist situates herself as a novice member of an unfamiliar community and, through engaging in the daily lives of the individuals around her, attempts to acquire an acceptable level of cultural competence. The actions, rituals, practices and behaviors of the natives, as well as their firsthand accounts of what such social "things" mean, are entwined with the anthropologist's trajectory of inclusion to create a "thick description" of the community under study (Geertz 1973). The product of this effort is *ethnography*, literally "the writing of culture."

The emphasis on locality and individuals in anthropology exists in some tension, albeit of the complementary kind, with communication scholarship, which looks to discourse as the primary autopoetic mechanism through which nuclear institutions reproduce themselves (Taylor, Kinsella, Depoe, and Metzler 2005). Communication scholars study the discursive elements of human society, insofar as the signs, codes, words and images produced and consumed by people and the institutions they inhabit provide material for scholarly analysis. However, between communication and anthropology there is considerable overlap in subject areas, interests, philosophical frameworks, and even methods. These days, it is not unusual for anthropologists to incorporate the study of discourse in their fieldwork, while communication scholars use ethnographic methods to study the discourse of cultural groups. Moreover, communication and anthropology recognize what Michel Foucault made most explicit: that discursively defining the nature of truth necessarily involves the exercise of power, such that knowledge and power are intimately intertwined (Foucault 1980; Kinsella 1999). Anthropology's critique of truth thus lies in what Don Donham (1992) calls its "critical moment," its ability to illuminate the constructed nature of the taken-for-granted through exploring the exotic. Communication seems more directly focused on discourse, dissecting the rhetorical frames and techniques through which power is exercised in the forms and practices of truth-telling. In many ways, communication and anthropology bring complementary perspectives to the critique of knowledge, although the potential is perhaps not realized as frequently as it could be.

This is certainly true in regard to nuclear weapons. As détente fell apart in the late 1970s, communication scholars engaged in an intense period of studying, analyzing, and critiquing the contentious, oftentimes paradoxical discourse swirling around nuclear technologies (Taylor 1998). The field has produced a rich body of social critique, particularly when focused on problems of social control, reproduction, power and meaning in complex societies—all themes that dominate nuclear weapons-related discourse. In contrast, anthropologists remained largely silent about such themes as nuclear war, deterrence, and realist theories of international relations (see especially Weldes, Laffey, Gusterson and Duvall 1999). Gusterson (1999) has criticized international relations theorists for being so caught up in the dominant logics of security studies discourse that they missed the end of the Cold War. A similar critique could be aimed at anthropologists, for whom the Cold War's end "barely registered" as an ethnographic problem (Weldes, Laffey, Gusterson and Duvall 1999, 6).

Yet in the wake of the Cold War's end, a handful of anthropologists who came of age during the height of the Cold War began to pursue research in, around, and for the nuclear weapons laboratories. This trend is largely attributable to the three-decade repatriation of the discipline that began in the 1970s (Fox 1991; Hymes 1974), which gradually increased the intellectual credibility of research in the United States vis-à-vis more traditional ethnographic excursions abroad—a problem, I might add, that communication scholars never had to address. As a result, anthropologists discovered some very interesting ethnographic problems in their own backyards (see especially Ortner 1991). Hugh Gusterson's 1996 book about Lawrence Livermore National Laboratory was a breakthrough ethnography; not because Gusterson was the first social scientist to study a site in the US nuclear weapons complex, but because it was the first anthropological monograph to seriously apply frames and theories usually reserved for the nonwestern Other to a group of Western politico-scientific elites. Since then, a few anthropologists have pursued fieldwork about the nuclear weapons laboratories (for example, McNamara 2001;

Gusterson 2004; Masco 2006). Yet for many reasons—access to secretive institutions and funding being two of the most important—such work remains relatively rare in the anthropological community.

Knowledge as Construction, Communication, Culture

Treating science as a cultural problem is a relatively new phenomenon in anthropology, and there are not many truly anthropological studies of Western scientists (with a few notable exceptions; e.g., Dubinskas 1988; Gusterson 1996; Traweek 1988a, 1988b, 1992, 1996). For the most part, anthropologists have left the study of Western scientific institutions to historians, philosophers and sociologists, while choosing instead to make a scientific problem out of non-Western cultures. Outside anthropology, however, there are many ethnographic studies of Western science, most of which emerged from the sociology of scientific knowledge (SSK) movement that developed in France and Great Britain during the 1970s and 1980s. Until the 1970s, most sociology of science in the United States and Europe reflected the institutional emphasis of the Mertonian school. Then, "without much anthropological involvement," writes Bryan Pfaffenberger (1992, 491), European sociologists "discovered" participant observation as research tool, producing a steady stream of laboratory-based ethnographies. Influenced by the writings of Thomas Kuhn and Michael Polanyi, and using ethnomethodological approaches, they began to study the social settings and micro-interactions generative of scientific knowledge (e.g., Latour 1987; Latour and Woolgar 1986). Sociologists in this movement sought to "strip science of its extravagant claim to authority" by demonstrating the significant role that social negotiation plays in the production of scientific knowledge (Callon and Latour 1992, 346): science viewed through the lens of Foucauldian critique (e.g, Latour 1987, 182–183; see also Kinsella 1999).

This body of work speaks to communication scholars and anthropologists alike: the former because it emphasizes science as a bundle of complicated and evolving communicative practices (Kinsella 1999, 174); the latter because it uses ethnographic frames and methods to illuminate science as a social phenomenon. However, where the social studies of science tends to be tightly focused on the nature of scientific knowledge, communication and anthropology ask broader questions about how communities—scientific or otherwise—constitute themselves. As I discuss below, a nuanced understanding of science as a cultural phenomenon requires that we move beyond questions of technical epistemology to focus on issues of community and identity, if we are to understand the remarkable resilience of scientific knowledge as a form of human practice.

An Ethnographic Encounter with Knowledge Loss

As an anthropologist, I spent six years as a participant-observer and researcher at the Los Alamos National Laboratory, conducting *in situ* observations of nuclear weapons scientists, technicians and engineers. I came to Los Alamos as a doctoral student in 1997 to study the emergence of "diversity" discourse in the laboratory workplace. To my surprise, I found that my identity as an anthropologist often produced jokes that played on tensions engendered by aging and inactivity among experts whose work was most deeply impacted by the end of testing. One afternoon, a physicist I lunched with commented, wryly, "I guess I can see why you'd want to study us. We're becoming a bunch of relics." Later, a middle-aged engineer stared at me when my physicist friend described me as an anthropologist studying the laboratory: "Don't you folks usually study dinosaurs?" he asked, apparently mistaking anthropology for paleontology. Then he looked at the physicist and said, with a slightly sarcastic laugh, "Wait a minute. I keep forgetting that we *are* dinosaurs."

Throughout the laboratory I met scientists who expressed concern that their knowledge was not as valued as it once had been, despite the fact that the laboratory's senior managers were emphasizing the importance of preserving and transferring certain skills and abilities. This point was dramatically illustrated when I visited a senior engineer whose office was decorated with a remarkable collage of nuclear ephemera: small posters, t-shirts, hats, stickers, certificates, and photographs from twenty years of nuclear tests. He explained to me that experienced weapons engineers were talking about retirement and, given the laboratory's foundering mission, he found it difficult to entice new staff with the promise of challenging work. He missed the Nevada Test Site and worried that younger staff members who lacked NTS experience might not be able to execute a weapon test if asked to do so. "Don't you anthropologists work with Native Americans to preserve stories, art, legends?" He paused and looked out the window, then looked back at me. "I mean, how do you save a dying culture?" That evening, driving back to my little house in nearby Pojoaque, I passed Black Mesa, a volcanic formation that rises above the Rio Grande Valley and has deep historical significance for the Native American people of the nearby Pueblos. Nuclear weapons, cultural survival, engineering, and indigenous knowledge, I thought. What could they possibly have in common?

Over time, I came to realize that the ideational connection between these two is *identity*. Identity is more than the social categories in which we claim membership; it is our way of being in the world, as expressed in the countless actions we take throughout the course of a day, a year, our lives. Identity is simultaneously individual and collective, an emergent property of the many millions of interactions that people have with each other and the world around them. Studying identity requires observing people as they move through the world, as well as listening as they verbalize their understanding of the world. Identity can often be found as the predicate in the sentences through which we locate ourselves in the world *and* ascribe a location to others: I am a guitar player; we have a band; you are not from around here; my parents are Catholic. Statements like these not only verbally instantiate our *selves* vis-à-vis the rest of the world; they comprise texts that can be parsed, read, studied to understand the dynamics of emergent and evolving identity. Texts are the sensible traces of our ways of being and knowing in the world.

In this regard, community of practice theory (Lave and Wenger 1991; Wenger 1998) provides a framework capable of integrating communication's focus on discourse with anthropology's emphasis on culture. Communities of practice are variably organized entities that emerge over time as individuals engage with each other, and with various aspects of the physical world, in the sustained pursuit of a particular enterprise (Wenger 1998, 45). They are the geographical and temporal "places" in which cultural forms emerge and are perpetuated through the actions, beliefs, rituals, behaviors of people. Individuals belong simultaneously to multiple and overlapping communities of practice, which vary in size, formality and level of integration, from the family, to the classroom, to the tribe, to the corporation. As "purposive sets of relations . . . among persons, activity, and the world," communities of practice "are an intrinsic condition for the existence of knowledge," because they provide interpretive frames of reference that make human action meaningful (Lave and Wenger 1991, 98). As such, community of practice theory broadens the discussion of tacit-versus-explicit-knowledge in science by linking the dynamics of knowing to the ongoing reinscription of shared identity, as expressed in a community's discursive products and through the practices of its members.

Where the sociology of science is largely silent on the significant emotive issues that lie behind jokes and comments about "dinosaurs," "relics," and "dying cultures," community of practice theory requires that we ask questions about the discursive perpetuation of worldviews, and the relationship between worldview and the loss of tacit knowledge. Understanding these requires not only making as visible as possible the cognitive processes and negotiations that generate knowledge in some fixed form, but also accounting for the social, political, economic, local and global contexts that make cognitive activities meaningful. In looking to see how the global makes its presence felt, however subtly, in the micro-ways that people live their work, anthropology recontextualizes the cognitive processes generative of scientific and engineering knowledge. Only in doing so can we fully appreciate the extent to which scientific pursuits are, like all other ways of knowing, inherently local, time and context dependent activities.

THE SOCIAL PRODUCTION OF NUCLEAR CONFIDENCE

Weapons science is a hybrid of engineering and physics, an applied discipline that focuses on a few microseconds of transition when firing energy from a fuse or trigger enters a stable nuclear explosive system and causes its parts to move, compress, merge, and finally blow apart (Hoddeson, Henriksen, Meade and Westfall 1993). Roger Shattuck has described this enterprise as a form of "forbidden knowledge," an intellectual endeavor that "takes place on a slippery slope between pure knowledge and its application in the real world" (1996, 182). The ability to manipulate this knowledge, to transform the intimate understanding of arcane physics principles into working prototypes for new nuclear explosives, has been the preoccupation of several generations of the laboratory's nuclear weapons community. Thousands of physicists, technicians, chemists, engineers, and other technical experts have devoted decades of career time, spent billions of dollars, blown up entire islands, and contaminated hundreds of square miles of Nevada desert in an effort to explore and characterize the nanosecond dynamics of nuclear explosions (Schwartz 1998).

For fifty years, the driving principle behind this vast effort was nuclear deterrence, technologically instantiated in the strategic triad of land, air and sea-based missiles and bombs equipped with "physics packages"—working nuclear explosive systems designed at Los Alamos, Livermore and Sandia, and mass-produced in the DOE's manufacturing complex. The resulting knowledge about the safety, security, and reliability of nuclear explosives provided the foundation for nuclear confidence, the principle that transformed the recursive, dead-end logic of nuclear deterrence into a workable foundation for defense policy (Rosenthal 1990). As one of the LANL's senior policy analysts explained,

> The heart and soul of any successful policy of mutual nuclear deterrence is the certain belief of national leaders, beyond reasonable doubt, that their own and their adversaries' nuclear forces are . . . deliverable and will function as intended under any circumstances . . . [this belief] rests solely on the assurances given to those leaders by scientists, and by the credibility that those scientists have with the leaders (White 1987a, 2; see also White 1987b).

Or, in the words of former LANL director Sig Hecker, "the credibility of the U.S. nuclear deterrent policy rests indispensably upon the credibility of the three DOE nuclear weapons laboratories" (1988, 4–6; also see Fetter 1988).

Nuclear testing provided an epistemological basis for nuclear confidence, but not in a classical statistical sense. Although tests provided a great deal of data about explosive performance, they were far too expensive and difficult to perform multiple trials for any weapon system, much less isolate and repeatedly measure a single feature of a primary or a secondary. Moreover, the exigencies of Cold War weapon development required the weapons laboratories to focus on the next weapon system, not generate a statistically significant number of samples for established designs. Instead, American military strategists and war planners—the consumers of the laboratory's products—trusted the collective expertise of the laboratory's weapons experts because of their long track record of successfully designing, engineering and detonating a wide variety of working nuclear explosives. In a very real sense, the credibility of the nation's nuclear deterrent was rooted in the expertise of the individuals with the most intimate knowledge of nuclear explosives, so that the laboratory's weapons-related judgments were as much the bedrock of nuclear deterrence as were the weapons themselves.

The Design and Test Cycle and the Social Organization of Nuclear Confidence

Understanding how the production of confidence has changed in the past decade requires understanding the experimental cycles that once structured work at the nuclear weapons laboratories. Throughout the Cold War, interactions between the military and the DOE's weapons facilities were coordinated through an eight-phase acquisition cycle that structured the design, development, testing, manufacturing, stockpiling and retirement of nuclear weapon systems. Within this cycle, nuclear weapon designers at Los Alamos[1] played a key role as the voice of confidence. Housed in the evocatively named X Division, primary and secondary designers and their counterparts in LANL's weapons engineering groups worked closely with the military to identify advanced concepts for new nuclear weapons and developed experiments to validate novel designs. As such, this small coterie wielded enormous power in outlining the research agenda for the weapons programs.

However, designers neither built their own devices nor fielded their own tests. Instead, they relied on a vast, multidisciplinary community of scientists and engineers to translate design concepts into a working set of functional experimental artifacts.[2] When X Division was ready to field an experiment, it issued a design release that mobilized workers at Los Alamos and at NTS to begin preparations for a nuclear test. This could take several years of effort, as a typical nuclear test required the team to conduct preparatory high-explosive experiments, select a hole in Nevada, develop diagnostics for data collection, create a security plan and review all safety requirements, build the experimental device itself, design and machine a rack to hold the experiment and all associated equipment, lower the rack into the ground, backfill the hole, and finally detonate the device. A very complicated test could easily cost tens of millions of dollars and involve hundreds of staff members: weapon designers, diagnostic physicists, machinists, secretaries and other administrative assistants, physical security experts, mechanical and electrical engineers, radiochemists, construction engineers, drillers and mining experts, engineering technicians, materials scientists, geologists, crane operators, and electricians, among other disciplines.

Given the complexity of the testing program, I can safely say that no single individual ever understood everything there was to know about conducting a nuclear test. Instead, the design and test cycle was a shared *activity system* that integrated social, material, and individual components in the weapons programs, creating a context in which human agents could engage each other meaningfully in a collective problem-solving process (Keller and Keller 1996, 126). Every test catalyzed the emergence of a novel configuration of experts, each of whom understood their location in relation to their peers, and took responsibility for their role in bringing the experiment to fruition. Novice weaponeers learned their responsibilities by engaging with a particular sub-community of experts, with the level of responsibility, risk, and exposure increasing in proportion to the learner's level of participation in the process. This experimental program enabled its participants to reproduce and extend the social practices of nuclear confidence, and the collective identity of an integrated community, across generations of experts and weapons.

The design-and-test cycle waxed and waned but never really stopped. There was always another test on the horizon, iteration after iteration, like a series of waves, slowly rising and building toward an end point, then breaking into memory to make space for the next event. The many groundshaking nuclear tests that Los Alamos conducted demonstrated that both the weapons and their creators "worked," that weapons experts could reliably be expected to produce functioning devices and detonate them without mishap in the Nevada desert (Gusterson 1996; Pinch 1991; see also Collins and Pinch 1998).

Testing and the Production of Nuclear Truth

During debates over nuclear test restrictions in the late 1980s, one of the laboratory's senior weapons engineers was known for pointing to a map of Nevada and exhorting his fellow weapons experts, emphatically, to remember that "TRUTH is generated HERE." The use of the word "truth" evokes Michel Foucault's famous characterization of truth as "linked in a circular relation with systems of power that produce and sustain it, and to the effects of power which it induces and which it extends" (Foucault 1980, 133).

In his study of the Princeton Plasma Physics Laboratory, Kinsella uses Foucault's proposition to demonstrate that scientific facts are the product of negotiations that occur within an institutional matrix that structures the social and cognitive practice of scientists (1999). Similarly, in a place like Los Alamos, knowledge products are not confined to disciplinary boundaries, laboratory property, or administrative structures. As Hugh Gusterson (1996) points out, nuclear weapons are meaningful because they exist as technological reifications of the moral, social and political principles embedded in deterrence theory. Nuclear weapons are symbolically powerful because they carry messages about the just use of threat to prevent conflict; about the power of technology to curb the inherent violence of human society; about the Machiavellian rightness of means that effect a particular end. At LANL, scientific truths were (and remain) so tightly embedded in the political machinations of Cold War conflict as to be inseparable elements of the same historical context. The weapons experts who developed the stockpile were not simply makers of deadly machines; rather, their work actively reproduced and extended a regime of truth in which nuclear explosives exist as a redemptive technology, one whose very deadliness curbs what they believe is a "natural" human tendency towards violence.

All the weapons experts I met in my research, Cold War and post–Cold War alike, expressed a strong sense of individual commitment to maintaining a safe, secure, and reliable nuclear deterrent. My interviewees were proud of their efforts to design and certify reliable nuclear devices in support of the nation's nuclear deterrent. As one retired weapon physicist told me, emphatically, "Los Alamos has maintained a culture of quality. There is no industry in the United States that could afford the kind of quality, the guarantee that we give our weapons. We've always offered the damndest guarantee of weapon reliability." Although I was initially taken aback by statements like these, I learned not to automatically interpret them as evidence that the laboratory's weapons designers and engineers look forward to the day when their claims about the reliability of their devices will be vindicated in a nuclear war. On the contrary, the weapons community makes these claims loudly and clearly in the firm belief that nuclear confidence offers the best possible means of preventing conflict.

The mutual constitution of nuclear weapons, nuclear confidence and the subjectivities of weapons experts was dramatically illustrated for me during one of my visits to the Nevada Test Site in the spring of 1999, when I attended a tour that the laboratory had organized as part of a larger training exercise for novice weapon engineers. Our group stood around an abandoned test rack as the tour guide, a former NTS electrical engineer, explained the swift process of detonating a nuclear device. As his audience dispersed to explore the rusting equipment scattered around the event site, I stayed back and asked him if he missed working on nuclear tests.

"Testing?" he barked. "Of course I miss testing." He looked at me impatiently and gestured to the rest of the tour group. "Over here, folks, I'll show you the trailers where the arming and firing systems went." He started to walk away but I swung into step with him, my steel-toed safety boots pushing hard against the sand as I matched his long stride. "Why?" I asked. Without pausing, he turned around to look at me, took a few half-steps backward. "Why? Because it's a powerful thing, seeing a crater collapse into the ground." He turned forward again and continued talking more loudly, not looking at me but looking ahead, towards the horizon, as though he were talking to someone else. "I know what these things can do," he said, sounding almost frustrated. "I've seen them send a ripple a hundred feet high across the desert. Goddammit, I'd bring every world leader here if I could, I'd blow one up and make them watch that ripple. Just to show them. So they don't ever, ever forget what they're dealing with."[3] And he marched ahead of me, alone, shaking his head, the wind lifting thin gray strands of hair off his forehead and pushing his worn nylon jacket tightly across his barreled chest.

WEAPONEERING UNDER THE NEW PARADIGM: STOCKPILE STEWARDSHIP

For the nation's nuclear weapons laboratories, the Cold War went out with a bang. On September 23, 1992, at 3:04 in the afternoon, a small group of physicists and engineers from Los Alamos stood in an underground cement bunker at the Nevada Test Site and detonated an experimental nuclear device for a test called Divider. The previous month, Congress had passed the Hatfield amendment, ending the testing program after several dizzying years of change in national security policy. Between 1993 and 1995, Los Alamos went through a series of workforce reduction initiatives that cut the total size of its workforce from a high of roughly 15,600 employees to less than 12,500. Morale in the weapons programs reached a nadir as "the normal flow in and out of the workforce was seriously disrupted . . . over the four year period 1993–1996, [Los Alamos] hired a total of about 115 scientists and engineers while more than 400 departed" (United States Commission on Maintaining United States Nuclear Weapons Expertise 1999, 9). One prominent geophysicist, describing life at the laboratory in the early 1990s, said,

"The whole place was in free-fall. You know, people said, 'You'll never be able to sell your house, just leave it and walk. This place is collapsing, because nobody wants it, it doesn't have a role anymore, it doesn't have a mission'" (Chick Keller, quoted in Vasquez 1997, 69).

Ironically, the Clinton administration's opposition to nuclear testing instead brought a reversal of fortune to Los Alamos, since the establishment of Science Based Stockpile Stewardship (SBSS) meant renewed investment in experimental and computing facilities and experts at all three of the DOE's design agencies. By the time I arrived at Los Alamos in 1997, the laboratory had made a firm commitment to stewardship as the reigning mode of knowledge production. This commitment took shape in the "multiple forms of discourse, including oral, textual, and material productions" (Kinsella 1999, 175) through which Los Alamos instantiated itself as a post–Cold War institution capable maintaining its ties to the past as a hedge against uncertainty, while simultaneously embracing Science Based Stockpile Stewardship as the grand challenge of the future. For example, in 1999 testimony to the Senate Armed Services Committee, then-laboratory director John Browne stated,

> Maintaining the safety and reliability of our nuclear weapons without nuclear testing is an unprecedented technical challenge. The Stockpile Stewardship program is working successfully toward this goal, but it is a work in progress. . . . I am confident that a fully supported and sustained program will enable us to maintain America's nuclear deterrent without nuclear testing . . . The Stockpile Stewardship Program has undertaken this unprecedented technical challenge, and to date it is working (1999).

Within Los Alamos, the message that testing was a thing of the past, and SBSS the way of the future, was hard to escape: The new-employee orientation session I attended on my first day of work included a half-day session on the recent shift from forty-seven years of nuclear testing to SBSS. On a nearly daily basis, briefings, newsletters, white papers, electronic mail, journal articles, the LANL website, press releases, and the like provided opportunities for different parts of the nuclear weapons programs to describe their progress in making stewardship a success. In three years of fieldwork, my collection of Laboratory-generated, stockpile-stewardship-related ephemera grew to fill several large file drawers.

Yet institutional commitment notwithstanding, the shift from testing to stewardship was more contested than official discourse might suggest. Even with a revitalized mission and generous funding, weapons experts have spent the past decade struggling to make sense of ambiguous territory. The circular paradox of post–Cold War national security—in which nuclear confidence must be maintained without testing under a CTBT regime that itself depends, at least in part, on the DOE's success in maintaining the nuclear deterrent under Stockpile Stewardship—is tenable only because the weapons laboratories remain trusted stewards of the stockpile.

For weapons experts charged with making this policy work, knowledge loss is a very real problem. I often heard them ask the same questions raised by Mackenzie and Spinardi: Will the laboratory be able to conduct a test, should a future President order withdrawal from the CTBT? Without test experience, will future generations of weapons experts understand critical weapons problems? What about their credibility? As Robert, a senior primary designer at Los Alamos, lamented,

> For reasons I don't entirely understand, weapons design never had the hallmarks of a true profession. Lawyers have their bar exam, doctors have medical boards, but we don't have anything like that. Why? Because it was understood that the important people were tested by nuclear test experience. . . . Now who certifies the experts, and in the future, who will certify their replacements?

Reproducing this trust across generations *and* paradigms, without the epistemological bridge of nuclear testing, has created a great deal of stress for weapons experts, who often speak of a race against time to develop, verify and validate the SBSS toolkit and its practitioners, lest a surprise catch the community unprepared and force withdrawal from the CTBT.

The difficulty of keeping a foot in the past, racing into the future, and simultaneously maintaining confidence in an aging stockpile was a theme not entirely absent from official discourse around stewardship. For example, this excerpt from an annual report describing LANL's stockpile surveillance activities characterizes the past as a time of confidence, and the future as a period of uncertainty:

> In the past, our mission was accomplished on a large scale with growth. Stockpile systems were periodically replaced with newer and better versions, a robust design and production capacity supported both stockpile modernization, and the rapid implementation of stockpile repairs, and confidence was assured with the certainty of an underground nuclear test. [But] current plans require systems to remain in the stockpile indefinitely, and therefore confidence in the readiness of the stockpile now includes an uncertainty driven principally by aging . . . changes resulting from aging are expected from fundamental properties . . . aging mechanisms that cause these potential changes include the in-growth of decay products, damage, and associated void formation (LANL 1997, 1).

Moreover, without the design and testing program as a mechanism for cultural and technological renewal, themes of age and death emerge as worrisome and salient fears not just for the stockpile, but for its stewards. As the

DOE's Office of Defense Programs noted in an early description of stockpile stewardship,

> In the past, continuous development and production of new weapons maintained the scientific and technical knowledge and skill base essential for maintaining the safety and reliability of the stockpile. With no new weapons in development or production, budget reductions, and an aging staff with actual experience in designing, testing and producing nuclear weapons, the knowledge and skill base unique to nuclear weapons will atrophy (DOE-ODP 1995, 5).

Dealing with the problem of an aging stockpile was a relatively straightforward problem: In the 1990s, the DOE introduced the Stockpile Lifetime Extension Program, or SLEP, which consists of a series of Lifetime Extension Programs, or LEPs one for each system remaining in the nuclear stockpile. Each LEP is designed to refurbish the warhead with updated components and subsystems, thereby extending the design lifetime of each individual system and the stockpile as a whole.

While refurbishing a warhead is one thing, maintaining the collective expertise of the Cold War weapons community is another problem entirely. As I describe below, Los Alamos and its sister design agencies addressed the problem of knowledge loss by developing strategies to capture, preserve, and transfer knowledge as a hedge against an uncertain future. Nevertheless, many senior weaponeers were frustrated with what they perceived as a lack of effort on the part of DOE to adequately address the problem of knowledge loss. Indeed, in 1998, the Congressionally-mandated Commission on Maintaining United States Nuclear Weapons Expertise (also known as the Chiles Commission) released a substantial report criticizing the DOE for its failure to develop a comprehensive knowledge management and expert succession plan for the weapons complex (United States Commission on Maintaining United States Nuclear Weapons Expertise 1998).

Some of this frustration is attributable to the shaky epistemological assumptions that underlie knowledge preservation: as I discuss below, the entire project rests on a misleading metaphor of knowledge as a commodity. However, much of it is directly related to stockpile stewardship itself, which represents a dramatic shift in the social organization of the nuclear weapons programs.

Knowledge Loss, Knowledge Preservation, and Impact Constituencies

Post–Cold War knowledge preservation efforts at the national laboratories tend to fall into three categories: the collection and archiving of artifacts generated during the Cold War (drawings, memos, reports, input decks for computational models, physical mock-ups of engineered devices, and the like); the elicitation of Cold War expertise from an individual or a small group of individuals whose knowledge is considered unique; and formal training programs designed to "transfer" skills across the generation gap.

In the early 1990s, Los Alamos established the Nuclear Weapons Archiving Program, or NWAP, to coordinate and fund a variety of archiving projects, including Cooperative Research and Development Agreements (CRADAs) with major industrial partners like IBM and Xerox to develop electronic archives, sophisticated scanning and retrieval procedures, and computer-based knowledge management programs. Similarly, in 1993, Livermore established the Nuclear Weapons Information Project (NWIP), an effort to capture at-risk knowledge for transfer to future scientists and engineers. A year later, NWIP morphed into NWIG, the Nuclear Weapons Information Group, which included representatives from across the DOE, DOD, and even the UK's Atomic Weapons Establishment (AWE). The goal of NWIG was to establish a broad information preservation effort across all the sites that had historically collaborated with each other in the design, testing, production and dismantlement of nuclear weapons. At Sandia National Laboratories, the NWIG took the form of a Knowledge Preservation Project, or KPP, that brought together groups of weapons experts to be videotaped as they discussed their involvement in specific programs and projects. The KPP staff also created transcripts and indices so that the video recordings comprise an electronically searchable database.

Formal training programs also emerged as a strategy to combat knowledge loss. In the late 1990s a group of mid-career and senior weapon designers in X Division formed the Theoretical Institute for Thermonuclear and Nuclear Studies, or TITANS, a two-year, formal, classified postdoctoral training program for novice weaponeers, complete with a peer-reviewed thesis and oral defense with senior weapon designers as committee members. Similarly, a small group of test-trained weapons experts at Sandia National Laboratories instituted a formal Weapons Intern Program in which a dozen or so junior-level engineers from the Sandia workforce would spend a year immersed in learning about Sandia's role as the engineering link between the DOE and the DOD. Sandia's program was formally accredited in 2001, when it entered into a Memorandum of Understanding with the New Mexico Institute of Mining and Technology in Socorro to provide students twenty-one graduate credit hours toward a master's degree in energetic materials.

Although these programs indicate substantial momentum within all the laboratories' workforce around the issue of knowledge loss, none of the nation's three design laboratories ever made a top-down, comprehensive commitment to full institutional knowledge capture, management, preservation, or transfer programs, despite the fact that LANL managers often publicly commented on the value of greybeard expertise and the importance of capturing it for the future. Several factors explain the uneven commitment to knowledge preservation: For one thing, not everyone agrees that maintaining full design and test expertise is necessary for stockpile stewardship to succeed (see especially McKinzie, Cochran and Paine 1998). Moreover, training programs have been fiercely criticized as efforts to subvert the intentions of the CTBT. LANL's sister institution, Sandia, came under fire in 1998, when it was revealed that engineers were planning a flight test of the Bomb Impact Optimization System, or BIOS, a guidance system to improve target acquisition. Despite Sandia management assurances that the engineering design project was intended as a hands-on "exercise to hone the skills of Sandia's weapon design [engineers]," the laboratory was accused of attempting to get around Congressional restrictions on weapon design activities, and the program was abruptly canceled (Fleck 1998). Electronic knowledge management programs were weakened by Los Alamos' much publicized security problems, which caused Congress, the DOE, and external critics to focus on computer and electronic security at the laboratories. In 1999 the *Bulletin of the Atomic Scientists* published a piece entitled "Steal This!" alleging the laboratories were developing "the most attractive nuclear espionage target ever developed . . . a point-and-click computer network of weapons knowledge so complete that its theft by foreign spies would constitute a loss of virtually every nuclear weapon design secret possessed by the United States." Not surprisingly, the database that the *Bulletin* warned about never came into existence (Stober 1999).

Far and away, however, the biggest reason for the laboratory's uneven commitment to knowledge preservation was its continuing mission vis-à-vis the nuclear stockpile. Los Alamos and its sister laboratories were and remain weapons development facilities, not archives or museums, and most of my colleagues around me were far more concerned about maintaining the stockpile in the here-and-now than they were about the eventuality of knowledge loss. Stockpile stewardship both aggravates and mitigates the knowledge loss problem, because the program represents a fundamental shift in the laboratory's traditional mode of producing knowledge: as former LANL Director Browne described it,

> The United States developed its nuclear arsenal using the same methods applied to most other complex systems: a sequence of design-test-produce. . . . Today we are employing a new method: a sequence of surveillance-evaluation-response. In this new paradigm, we are using a fundamentally different set of tools to ensure the safety, reliability, and performance of nuclear weapons: the Stockpile Stewardship Program (Browne 1999, 4).

One consequence of this change is the displacement of Cold War experts whose traditional knowledge is no longer directly relevant to the work currently performed. For these individuals, knowledge preservation can be interpreted as a response to a rapidly shifting world. External disruptions bringing uncontrollable change are frightening, particularly when people perceive that traditional ways of knowing, markers of collective identity, are in imminent danger of disappearance. In response, people tend to become more conscious and deliberate about reproducing what is familiar. Although formal knowledge preservation and management programs are the kind of adaptive strategy that one might expect from a large, modern, Western, technoscientific bureaucracy like Los Alamos, anthropological studies of indigenous peoples describe similar patterns of response. For example, in her historical study of the Barolong boo Ratshidi of South Africa, Jean Comaroff describes how European colonization repeatedly overturned the Tshidi social system during the eighteenth and nineteenth centuries. Yet despite constant evictions, illness, proselytizing missionaries, environmental stress, and sporadic warfare, Tshidi chiefdoms worked to reproduce their spatial, agricultural, political, and domestic arrangements wherever they settled (1985, 42).

However, knowledge preservation is not just an adaptive strategy; it is a political one as well. As Bryan Pfaffenberger writes, during periods of social and technological adjustment,

> impact constituencies—the people who lose when a new production process or artifact is introduced—engage in strategies to compensate for the loss of self esteem, social prestige and social power caused by the technology . . . a technological drama's statements and counterstatements draw upon a culture's root paradigms, its axioms about social life; in consequence, technological activities bring entrenched moral imperatives into prominence (1992, 506).

If nothing else, the sociotechnical drama that is laboratory's adjustment to the post–Cold War era reveals that nuclear confidence remains the laboratory's entrenched moral imperative. Weapons experts' worries about the disappearance of knowledge reflect far more than the epistemological conundrum created by the end of testing. Instead, they highlight the social fault lines created when SBSS replaced the practices of testing with a new portfolio of tools and techniques.

At Los Alamos, impact constituencies are particularly visible among those disciplines most closely aligned with the Nevada Test Site, such as diagnostic physics. While the laboratory was still testing, diagnostic teams developed sophisticated arrays of sensors for downhole experiments; indeed, their central role in the production of nuclear confidence made the diagnostics groups some of the most politically powerful entities in Los Alamos. However, the realignment to stockpile stewardship, and drastically reduced activities at NTS, meant that downhole diagnostics would play a more limited role in the future. As a result, although diagnostic physicists are still heavily involved in small-scale and subcritical experiments, recruiting has dropped precipitously since 1992. According to laboratory demographics, in 2000, the average age of all technical staff in the diagnostic physics division was forty-nine years of age; indeed, one of the diagnostics groups lost most of its members to retirement between 1996 and 2000.

Perceiving a lack of interest in maintaining their expertise, some senior diagnosticians began in the mid-1990s to pursue strategies to capture and preserve their own problem-solving methods, expert judgment, unwritten knowledge, and experimental processes. For example, one NTS expert was championing an archiving project dedicated to the last series of nuclear tests in 1992, to create a state-of-the-art description of underground testing that would map the organizational interfaces in the testing program, provide a schedule for important events, and identify key positions and their responsibilities in relation to executing a nuclear test, in preparation for the day that the United States might return to testing.

Other senior experts assumed responsibility for archiving their own knowledge. One afternoon, I spent a couple of hours watching a retiree carefully annotating data from a test conducted in the 1980s. He explained that many factors could influence data quality, but that someone without test experience might not know to account for them. "This is all in my head," he said. "It could be very hard to evaluate this data when I'm not around."

Champions of these internal knowledge preservation projects were emphatic about the fragility of design-and-test knowledge and its significance for the laboratory's future. Often they had to struggle for funding for their efforts. Few believed that the laboratory could maintain the stockpile indefinitely without testing; consequently, they envisioned a future in which the weapons programs would be unprepared to conduct a nuclear test when—not if—required to do so. They saw their archiving and training work as an important, if incomplete, means of shoring up the laboratory's eroding Cold War knowledge base. For these experts, stewardship pushed aside their ways of knowing, and their concerns about knowledge preservation can be interpreted as an attempt to assert the continued value of their knowing selves to the institution's future. Yet these same people routinely acknowledged that many of the technologies used at NTS were already becoming obsolete, that the laboratory's workforce was changing, that the DOE was committed to stewardship for the foreseeable future, and that the political consequences of returning to testing would be unimaginably high in any case.

But Does It Work?

Even if the national laboratories had fully committed themselves to a broad program of knowledge capture, preservation, storage, retrieval and transfer, the benefits of such an effort are dubious. Knowledge management is a problematic enterprise because it relies on a metaphor of knowledge as a commodity that can be elicited, captured, preserved, archived, stored. However, knowledge is not a commodity; it is an emergent property of human beings "coming to terms with actions and products that go beyond the already known" (Keller and Keller 1996, 127).

Knowledge preservation programs at the national laboratories tended to focus on creating and saving *things*: documents, drawings, even mock-ups of experimental assemblies. In community of practice theory, such items are reifications, or objects that individuals create to congeal the experience of knowing into thing-ness. Reifications are "evocative shortcuts [that represent] . . . the tip of an iceberg, which indicates larger contexts of significance realized in human practices" (Wegner 1998, 58, 61). Latour (1987, 227; 236–237) refers to this property as "immutable mobility," meaning a text that fixes the emergent knowledge in a transportable form: as a drawing, a proposal, a model, a set of equations.

Considering the communicative role of reifications in the production of collective knowledge sheds light on the problems inherent in efforts to "preserve" knowledge. Reifications are not ends in and of themselves; rather, their importance is as communicative vehicles that catalyze meaning-making. In critiquing knowledge management as a communicative strategy, Heaton and Taylor (2002) differentiate between *artifacts* (documents, drawings, even spoken statements) and *texts* by emphasizing the role of the latter as living media for communication. Artifacts are just things; they become textual only when people actively use them to negotiate meaning (Heaton and Taylor 2002, 222). Shorn from its embedding activity, an artifact ceases to be a text and becomes inert, meaningless, unless it is drawn upon to further catalyze activity.

Rooted in a metaphor of knowledge-as-commodity, knowledge preservation not surprisingly mistakes reifications for the activity of meaning making. In an activity system like the Cold War design and test cycle, artifacts like drawings, memos, and input decks for computer models were embedded in the process of knowing. A document that was generated during preparations for a nuclear test lost its embedding context once the test was completed; with the end of testing, that context is receding into an increasingly distant past. Archiving projects may enable future community members to revisit that document in new context—the writing of institutional history, for example, or to figure out how a particular stream of data was collected—but the consequence of the document for organizational sense-making has changed (Weick 1995). As such, knowledge preservation falls short insofar as it focuses on developing and distributing stocks of potentially useful artifacts; the real challenge is ensuring their intelligent deployment in new and emerging contexts (Jackson, Poole and Kuhn 2002, 245).

Interestingly, it is in regard to emerging contexts that knowledge transfer programs may be having their greatest impact, insofar as they reinforce a sense of connection between the laboratory's past and its future. As I discuss below, the laboratory's mission remains a potent motivator for cognitive action, and institutional commitment to nuclear deterrence is the most important bulwark against knowledge erosion. Indeed, when this moral commitment is coupled with uncertainty about the aging stockpile, the resulting tension becomes the setting for the emergence of creative new ways of knowing about weapons, as well as novel frames for communicating confidence in the nuclear stockpile.

Plus que ça change, plus c'est pareil?

Anthropologist Bob Simpson writes that "knowledge, like so much else in society, is socially distributed. . . . [In complex societies] individuals participate in a partial and inchoate project in which knowledge and access to knowledge are variably distributed and expressed" (Simpson 1997, 44). Simpson makes this observation in a discussion of ritual change among the Berava drummers of southern Sri Lanka, but it is quite apropos of LANL's weapons community. As I have discussed above, Cold War–era knowledge of nuclear weapons design was a communal project (i.e., a structured process of knowing), in which the activities of many individuals came together in an integrated whole that was far more than the sum of its parts. Participating in this system required individuals to understand their social location in relation to other community members, and to develop a meaningful relationship with their peers.

Seen in this light, knowledge capture and transfer are far less important for nuclear confidence than reconstituting an activity system that enables people to assume responsibility for the laboratory's mission and the stockpile. The challenge for Los Alamos and its sister laboratories in the post–Cold War era is not the cryogenic preservation of the design and test cycle—itself an impossible task—but the development of a new activity system that catalyzes meaningful interaction across individual, organizational, and disciplinary boundaries. In other words, rather than focusing on how and what forms of knowledge will be lost in the future, it may make more sense to ask in what form the weapons community is continuing to reproduce itself, Cold War's end notwithstanding.

For one thing, Los Alamos' *raison d'etre* is still national security, although security discourse has changed dramatically since the late 1980s. During the Cold War, the bipower rivalry between the Soviet Union and the United States provided the center of gravity for security discourse. In contrast, the end of the Cold War has multiplied the United States' list of perceived and potential enemies to include proliferant states like Iran and North Korea, an emerging China, a resurgent Russia, and—particularly since 9/11—non-state actors and terrorist networks. As former CIA Director James Woolsey said, "We have slain a large dragon, but we live now in a jungle filled with a bewildering variety of poisonous snakes" (Senate 1997, 1). In this world of threats, neither President Clinton nor President Bush has advocated complete disarmament, despite steady reductions in the size of the stockpile. As a result, Los Alamos and its sister laboratories continue to enjoy generous funding for the nuclear weapons mission, and the past decade has seen Los Alamos' weapons programs recruiting and welcoming more new staff than one might expect, given that the institution is no longer designing and testing nuclear weapons.

From the moment these new employees join the LANL workforce, they are immersed in a tightly rationalist discourse that reproduces this "culture of insecurity" in its members (Weldes et al 1999). As Joseph Masco writes, the laboratory's mission lifts nuclear weapons experts away from LANL's rural enclave and places them squarely in the center of debates about international security, stability and warfare (1999, 210). Security—the bundle of understandings and practices that protect classified information from improper release or compromise—plays an important role in creating and maintaining the sense of realpolitik that permeates Laboratory culture. At Los Alamos, neophytes must master the practices and understandings of secrecy if they are to become fully vested members of the laboratory's secret world, in which the institution's core mission resides. Along the way, novices go through a series of transformative rituals, including being granted a basic laboratory badge, spending time as uncleared personnel in exile from the classified world, undergoing a federal clearance investigation, and finally being granted a security clearance and full entry into the classified world of nuclear weapons (McNamara 2001). This process of learning is a process of becoming: as Hugh Gusterson has noted, secrecy is the "anvil upon which the identity of new weapons scientists [at Lawrence Livermore] is forged" (1996, 68). As they become more integrated into the core areas of the laboratory, novices learn that their affiliation with classified activities marks them as targets for the hostilities and desires of various enemies, and that they accordingly bear personal responsibility for the security of the American nation-state. In this sense, the formation of a secret, knowledgeable self at the level of the individual is simultaneously a community-building process that draws individual subjectivities together, forming a culture whose worldview is characterized by wariness, suspicion, and a preoccupation with international threat.

Within this institutional milieu, programs like TITANS assume an importance beyond their explicit goal of training young weapon designers in the technical intricacies of Cold War weapon design. Instead, TITANS provides participants with a tantalizing sense of what it meant to be a part of the design-and-test community, and enables them to envision how they might contribute to creating its post–Cold War future. In many of the TITANS lectures, senior weapons experts, some of whom had been working in the weapons programs since the 1960s, peppered technical lessons about primaries and secondaries with tales of heroism and defeat: designers whose efforts to make a particular concept work led to serendipitous discoveries in weapons physics; miserable and embarrassing failures; designers who took an intractable problem and brilliantly transformed it into a working device against the expectations of their peers. Hero tales like these are not just colorful anecdotes; instead, they offer novices a repertoire of concepts, ideas, jokes, symbols, practices and beliefs that allow them to generate "'on the fly' coordinated meanings that allow the [community] to proceed" (Wenger 1998, 84). Moreover, a shared sense of history is critical for learning, insofar as novices must grasp the past if they are to engage meaningfully with present goals and perpetuate the community into the future:

> Interacting with old-timers offers living examples of possible trajectories . . . in a community of practice, old-timers deliver the past and offer the future, in the forms of narratives and participation . . . the possibility of mutual engagement offers a way to enter these stories through one's own experience (1998, 156–157).

Likewise, among novices, active interest in past practices—including dedicated participation in a program like TITANS—signals the creation of a relationship that connects their knowing selves to a weapons community in transformation: its history, its present condition, and the possibilities for its future. Although a sense of connection with institutional mission is not sufficient for maintaining the weapons community's cognitive authority over things nuclear, it is a necessary condition for novices to take the laboratory's mission into the future and, in the process, to redefine what it means to be a weapon designer in a world without design work, without testing, and without a Soviet Union. Hence, while several of Gusterson's interviewees at Livermore described stockpile stewardship as boring—one even likened stockpile maintenance to "polishing turds"—such ennui is notably missing among most of the primary and secondary designers that I met and observed, novices as well as test-hardened experts.

At the time that I observed TITANS classes in 1999, one of the program's participants was a young secondary designer who spent a summer at Los Alamos while finishing his doctoral research in high-energy physics at an Ivy League university. During that visit, his mentor at the laboratory had given him some simple physics calculations to work ". . . on the back of an envelope," he told me. "You can get pretty far with pencil and paper." The experience was exciting. "Working on a nuclear explosion is like reading data from a star," he said. After he finished his Ph.D., he came to Los Alamos, joined X Division and started modeling explosion dynamics for secondaries. In his job he works closely with two of the laboratory's most respected senior designers, both of whom tell stories about the design process, the days of testing, the importance of intuition in code development. "There's a lore to doing things," he said, "Ways that things are done and ways they are not done. I'm learning the lore." This is a far cry from the scenario that motivated X Division's senior designers to create TITANS: a weapons community split into an elder generation concerned with the stockpile and a younger generation that considered weapons research passé.

UNCERTAINTY, CONFIDENCE, AND THE STOCKPILE OF THE FUTURE

While newcomers' dedication to maintaining a nuclear deterrent may be a necessary condition for successful cognitive engagement with the problems posed by a post–Cold War stockpile, it is not sufficient. Without testing, certification of nuclear primaries and secondaries is a tricky project: In contrast to the nonnuclear systems in a warhead, nearly all of which can be decomposed into testable components and subsystems,[4] the nuclear primary and secondary function as integrated wholes. Stockpile Stewardship has to do more than fund a collection of expensive stovepiped research and development programs; it must enable the community to integrate knowledge about artifacts that must function as integrated wholes, while demonstrating its continued cognitive authority around the stockpile.

Ironically, uncertainty *itself* is emerging as the discursive basis of post–Cold War nuclear confidence; or more specifically, the development of formal methods for representing, analyzing, and minimizing uncertainty around the safety, security, reliability and performance of a nuclear device. Since around 2000, weapons experts at both Los Alamos and Lawrence Livermore—and increasingly Sandia—have been working to define, bound, and make confident statements about their knowledge: what they know about the aging stockpile, what they do not know, and to assess how "known unknowns" and "unknown unknowns" might impact weapon performance, safety, security, and reliability. New approaches to uncertainty, and the mathematical and statistical language and operations that enable weapon designers to represent and assess uncertainty, are commonly grouped as "Quantifying Margins of Uncertainty," or QMU. A recent Livermore publication compared QMU to testing and referred to the former as a "better way of certifying the nuclear stockpile" because it drives the weapons community to identify and resolve gaps in their understanding of nuclear performance (LLNL 2004). Within Los Alamos, developing new ways to think about uncertainty represents a major research area: for example, an Uncertainty Quantification Working Group brings together physicists, engineers, computer scientists, mathematicians and statisticians to debate and discuss the philosophy and mathematics of uncertainty on a regular basis. Not surprisingly, LANL's statistical sciences group has risen in political prominence vis-à-vis other research groups as interest in QMU grows. Indeed, one of the secondary designers who was in the TITANS classes that I observed recently completed an internal sabbatical with LANL's statisticians, so that he would better understand probabilistic approaches to modeling uncertainty in nuclear weapons problems.

QMU is also gaining discursive ground outside the national laboratories as a formal methodology for nuclear weapon certification. In 2002, when the DOE renegotiated the University of California's contract to manage Los Alamos, the fiscal year 2004 performance objectives required both LANL and LLNL to refine and demonstrate an integrated QMU certification methodology (NNSA Livermore Site Office 2004; USDOE, 2). The National Nuclear Security Administration has called for development of QMU under one of the Stockpile Stewardship Science Campaigns, and has asked the laboratories to establish a working QMU methodology by 2010 (DOE Office of Defense Programs).

The ultimate test for QMU could come in the form of the proposed Reliable Replacement Warhead (RRW), to which Congress in 2006 appropriated NNSA $25 million for further study. The idea of RRW is to replace warheads in the Cold War stockpile with a device that builds in very large margins against failure—in that sense, RRW as an artifact will reify QMU's concepts of margins and uncertainties in its very physical design. The NNSA has characterized this as a shift from "certify what we have built"—which is what stockpile stewardship currently seeks to do—to "build what we can certify" (NNSA 2006). Indeed, some claim that maintaining the Cold War stockpile into an indefinite future will require an eventual return to testing; RRW, say its proponents, can be certified without testing, and therefore obviates any possibility that the United States would invoke Safeguard F (see especially Medalia 2006).

So, even without a return to testing, the weapons laboratories seem to be quite successfully reproducing their specialized ranks, although the nature of "expertise" in these areas may change dramatically as a next generation engages with past knowledge and new techniques to establish novel understandings about a changing nuclear stockpile. This is because survival of the weapons community in whatever form is as much a matter of sociopolitical context as it is scientific practice. It is difficult to envision a time when "the unleashing of the nuclear genie is so unlikely that threats of [nuclear] retaliation become unnecessary" (Turner 1997, 106), and therefore weapons experts, defense strategists, and politicians continue to consider Los Alamos necessary for national security. However, it is important to remember this context is one that the laboratory itself makes possible and meaningful through its research, despite the fact that, like Traweek's high energy physicists (1988), weapons experts are extremely reluctant to acknowledge their own agency in shaping the world. As Wolfgang Panofsky has observed, "ultimately, we can keep nuclear weapons from multiplying only if we can persuade nations that their national security is better served without those weapons" (Panofsky 1994, cited in Mackenzie and Spinardi 1995, 88; see also Bundy, Crowe and Drell 1993). Until political and military leaders take decisive steps to change the context that makes weapons knowledge so valuable, experts at Los Alamos will continue to pursue new ways of knowing nuclear weapons that, in turn, reinscribe a remarkably resilient regime of nuclear truth.

NOTES

1. The judgments of weapon designers and engineers at Lawrence Livermore and Sandia National Laboratories were (and remain) equally important. However, my discussion focuses on LANL simply because that was where I conducted the bulk of my fieldwork.

2. For more detailed descriptions of the United States design and test cycle, see Wolff 1984, Machen 1988, Gusterson 1995; Coolidge 1996; also Federation of American Scientists www.fas.org/nuke/guide/usa/nuclear/testing.htm. Shkolnik 2002 provides a thorough overview of cold War test operations at the Soviet Semipalatinsk site.

3. Harold Agnew, former director of Los Alamos National Laboratory, is frequently credited as the person who first suggested that all political leaders should witness a nuclear test, so that they would be aware of their responsibilities to the world.

4. Sandia National Laboratories in Albuquerque, New Mexico, is the agency responsible for most of the engineered, nonnuclear subsystems in a warhead. Even under the test moratorium, SNL engineers have been able to test many of the components for which they are responsible. However, fiscal constraints are limiting Sandia's ability to perform extensive tests on its subsystems, with the result that Sandia is itself looking to a non-test based methodology for subsystem certification.

BIBLIOGRAPHY

Birely, John H. 1987. *The issue of nuclear testing*. Los Alamos, N.M.: Center for National Security Studies, Los Alamos National Laboratory.

Brown, Paul S. 1986. *Nuclear weapons research and development and the role of nuclear testing*. Livermore, Calif.: Lawrence Livermore National Laboratory.

Browne, John. 1999. *Testimony of John C. Browne, Director, Los Alamos National Laboratory to Armed Services Committee of the United States Senate, October 7, 1999*, available at armedservices.senate.gov/statemnt/1999/991007jb.pdf (Accessed May 7, 2000).

Bundy, McGeorge, William J. Crowe, Jr., and Sydney D. Drell, Sydney. 1993. *Reducing nuclear danger: The road away from the brink.* Washington, D.C.: Brookings Institution Press.

Callon, Steve and Bruno Latour. 1992. "Don't throw the baby out with the bath school! A reply to Collins and Yearly." Pp. 343–368 in *Science as practice and culture*, ed. Andrew Pickering. Chicago: University of Chicago.

Cimbala, Stephen. 1998. *The past and future of nuclear deterrence*. Westport, Conn.: Praeger Publishers.

Collins, Harry M. 1992. *Changing order: Replication and induction in scientific practice.* Chicago: University of Chicago Press.

Collins, Harry and Pinch, Trevor. 1998. *The golem at large: What you should know about technology.* Cambridge: Cambridge University Press.

Comaroff, Jean. 1985. *Body of power, spirit of resistance: The culture and history of a south African people.* Chicago: University of Chicago Press.

Coolidge, Matthew. 1996. *The Nevada Test Site: A guide to America's nuclear proving ground*. Los Angeles, Calif.: Center for Land Use Interpretation.

Donham, Donald. 1999. *History, power, ideology: Central issues in Marxism and anthropology*, 2d ed. Berkeley: University of California Press.

Dubinskas, Frank. 1988. Cultural constructions: the many faces of time. Pp. 3–38 in *Making time: Ethnographies of high technology organizations*, ed. Frank Dubinskas. Philadelphia: Temple University Press.

Fox, Richard G. 1991. *Recapturing anthropology: Working in the present.* Santa Fe, N.M.: School of American Research.

Fetter, Steve. 1988. Ch. 2 in *Toward a comprehensive test ban*. Cambridge, Mass.: Ballinger Publishing Company.

Fleck, John. 1998. Sandia scraps guided nuclear bomb project. *Albuquerque Journal*, 14 August: p. 1.

Foucault, Michel 1972. *Power/knowledge: Selected interviews and other writings, 1972–1977*, ed. Colin Gordon. New York: Pantheon.

——. 1984. *The Foucault reader*, ed. Paul Rabinow. New York: Pantheon Books.

Geertz, Clifford. 1973. *The interpretation of cultures*. New York: Basic Books.

——. 2000. *Local knowledge: Further essays in interpretive anthropology.* New York: Basic Books.

Gusterson, Hugh. 1996. *Nuclear rites: a weapons laboratory at the end of the Cold War*. Berkeley: University of California Press.

———. 1999. "Missing the end of the Cold War in international security." Pp. 319–345 in Weldes, J., M. Laffey, H. Gusterson and R. Duvall, eds. *Cultures of insecurity: states, communities and the production of danger*. Minneapolis: University of Minnesota.

———. 2004. *People of the bomb: Portraits of America's nuclear weapons complex*. Minneapolis: University of Minnesota Press.

Heaton, Lorna and James R. Taylor. 2002. Knowledge management and professional work. *Management Communication Quarterly* 16: 210–236.

Hecker, Sigfried S. 1987. *Statement of Dr. Siegfried S. Hecker, Los Alamos National Laboratory before the Senate Armed Services Committee concerning TTBT and PNET Ratification, February 26, 1987*. Los Alamos, N.M.: Center for National Security Studies, Los Alamos National Laboratory.

———. 1988. *The role of the DOE weapons laboratories in a changing national security environment.* Prepared for the U.S. Congress, House Committee on Armed Services, Procurement and Military Nuclear Systems Subcommittee. Los Alamos, N.M.: Center for National Security Studies, Los Alamos National Laboratory.

———. 1990. *Los Alamos: Science making a difference for the 21st century.* Los Alamos, N.M.: Los Alamos National Laboratory.

———. 1991. *Los Alamos: Science serving the nation*. Los Alamos, N.M.: Los Alamos National Laboratory.

Hoddeson, L., P. Henriksen, R. Meade, C. Westfall, G. Bayme, and R. Hewle. 1993. *Critical assembly: A technical history of Los Alamos during the Oppenheimer Years, 1943–1945*. London: Cambridge University Press.

Hymes, Dell. 1974. *Reinventing anthropology*. New York: Vintage Press.

Jackson, Michele J., M. Scott Poole, and Timothy Kuhn. 2002. The social construction of technology in studies of the workplace. Pp. 236–253 in *The handbook of new media: Social shaping and consequences of ICTs*, ed. Leah Lievrouw and Sonia Livingstone. Thousand Oaks, Calif.: Sage.

Keller, Charles and Janet Dixon Keller. 1996. Thinking and acting with iron. Pp. 125–143 in *Understanding practice: Perspectives on activity and context*, ed. Seth Chaiklin and Jean Lave. Cambridge and New York: Cambridge University Press.

Kinsella, William J. 1999. Discourse, power, and knowledge in the management of "big science": The production of consensus in a nuclear fusion research laboratory. *Management Communication Quarterly* 13(2): 171–208.

Latour, Bruno. 1987. *Science in action*. Cambridge, Mass.: Harvard University Press.

Latour, Bruno and Steve Woolgar. 1986. *Laboratory life: The social construction of scientific facts*. Princeton, N.J.: Princeton University Press.

Los Alamos National Laboratory. 1976. *Los Alamos Scientific Laboratory weapon program* (ADWP-76-21). Los Alamos, N.M.: Los Alamos National Laboratory.

———. 1997. *Enhanced surveillance program: FY 1997 accomplishments* (LA-13363-PR). Los Alamos, N.M.: Los Alamos National Laboratory.

Lave, Jean. 1988. *Cognition in practice: Mind, mathematics and culture in everyday life*. Cambridge: Cambridge University Press.

Lave, Jean and Etienne Wenger. 1991. *Situated learning: Legitimate peripheral participation*. Cambridge, UK: Cambridge University Press.

Lemert, Charles. 2001. *Social things: An introduction to the sociological life*. Lanham, Md.: Rowman & Littlefield.

Lukes, Steven. 1982. Relativism in its place. Pp. 261–305 in *Rationailty and relativism*, ed. Martin Hollis and Steven Lukes. Oxford: Basil Blackwell.

Machen, Judy. 1988. *The Los Alamos nuclear test program: Field test operations*. Los Alamos, N.M.: Los Alamos National Laboratory.

MacKenzie, Donald and Graham Spinardi. 1995. Tacit knowledge, weapons design, and the uninvention of nuclear weapons. *American Journal of Sociology* 101 (1): 44–99.

Marcus, George. 1999. "Foreword." Pp. vii–xv in *Cultures of insecurity: States, communities and the production of danger*, ed. J. Weldes, M. Laffey, H. Gusterson, and R. Duvall. Minneapolis: University of Minnesota.

Masco, Joseph. 1999. States of insecurity: Plutonium and post–Cold War anxiety in New Mexico, 1992–1996. Pp. 203–231 in *Cultures of insecurity: States, communities and the production of danger*, ed. J. Weldes, M. Laffey, H. Gusterson, and R. Duvall. Minneapolis: University of Minnesota.

——. 2006. *Nuclear borderlands: The Manhattan project in post–Cold War New Mexico.* Princeton, NJ: Princeton University Press.

McNamara, Laura A. 2001. *Ways of knowing about weapons: The Cold War's end at Los Alamos national laboratory.* Ph.D. thesis, Department of Anthropology, University of New Mexico.

McKinzie, Matthew G., Thomas B. Cochran, and Christopher E. Paine. 1998. *Explosive alliances: Nuclear weapons simulation research at American universities*. Washington, D.C.: Natural Resources Defense Council.

Medalia, Jonathan. 2006. *Nuclear weapons: The Reliable Replacement Warhead program*. Washington, D.C.: U.S. Library of Congress, Congressional Research Service.

Nader, Laura. 1972. Up the anthropologist—Perspectives gained from studying up. Pp. 285–311 in *Reinventing anthropology*, ed. Dell Hymes. New York: Pantheon Press.

Nolan, Janne. 2000. Preparing for the 2001 nuclear posture review. *Arms Control Today* 30, November: 10–14.

Ortner, Sherry B. 1991. Reading America: Preliminary notes on class and culture. Pp. 163–189 in *Recapturing anthropology: working in the present*, ed. Richard G. Fox. Santa Fe, N.M.: School of American Research.

Pfaffenberger, Brian. 1992. Social anthropology of technology. *Annual review of anthropology* 21: 491–516.

Pinch, Trevor. Testing—one, two, three . . . testing: Towards a sociology of testing. *Science, Technology and Human Values* 18: 25–41.

Rosenthal, Debra. 1990. *At the heart of the bomb: The dangerous allure of weapons work*. Reading, Mass.: Addison-Wesley.

Schell, Jonathan. 2000. The folly of arms control. *Foreign Affairs* 79 (5): 22–46.

Schwartz, Stephen, ed. 1998. *Atomic audit: The costs and consequences of U.S. nuclear weapons since 1940*. Washington, D.C.: Brookings Institution Press.

Sharp, David H., Timothy C. Wallstrom, and Merri Wood-Schultz. 2004. Physics package certification: From ONE to 1.0. Pp. 204–205 in *Theoretical division nuclear weapon program highlights 2004–2005.* Los Alamos, N.M.: Los Alamos National Laboratory.

Shattuck, Roger. 1996. *Forbidden knowledge: From Prometheus to pornography.* New York: St. Martin's Press.

Shkolnik, Vladimir S. 2002. *The Semipalatinsk test site: Creation, operation and conversion.* (Paul B. Gallagher, trans.) Albuquerque, N.M.: Sandia National Laboratories.

Simpson, Bob. 1997. Possession, dispossession and the social distribution of knowledge among Sri Lankan ritual specialists. *Journal of the Royal Anthropological Institute* 3: 43–59.

Stober, Daniel. 1999. Nuclear secrets: Steal this! *Bulletin of the Atomic Scientists* 55 (4): 14–16.

Taylor, Bryan. 1998. Nuclear weapons and communication studies: A review essay. *Western Journal of Communication* 62: 300–315.

Taylor, Bryan C., William J. Kinsella, Stephen P. Depoe, and Maribeth S. Metzler. 2005. Nuclear legacies: Communication, controversy, and the U.S. nuclear weapons production complex. Pp. 363–409 in *Communication Yearbook 29*, ed. Pamela Kalbfleisch. Mahwah, N.J.: Lawrence Erlbaum Associates.

Traweek, Sharon. 1988a. *Beamtimes and lifetimes: the world of high energy physicists.* Cambridge, Mass.: Harvard University Press.

———. 1988b. Discovering machines: nature in the age of its mechanical reproduction. Pp. 39–91 in *Making time: Ethnographies of high technology organizations*, ed. Frank Dubinskas. Philadelphia: Temple University Press.

———. 1992. Border crossings: Narrative strategies in science studies and among physicists in Tsukuba Science City, Japan. Pp. 429–465 in Andrew Pickering, ed. *Science as practice and culture*. Chicago: University of Chicago.

———. 1996. *Kokusaika, gaiatsu* and *bachigai*: Japanese physicists' strategies for moving into the international political economy of science. Pp. 160–173 in *Naked science: Anthropological inquiry into boundaries, power and knowledge*, ed. Laura Nader. New York: Routledge.

U.S. Commission on Maintaining U.S. Nuclear Weapons Expertise. 1999. *Report of the commission on maintaining United States nuclear weapons expertise: a report to the Congress and Secretary of Energy, pursuant to the National Defense Authorization Acts of 1997 and 1998.* Washington, D.C.: Commission on Maintaining U.S. Nuclear Weapons Expertise, 1999, www.dp.doe.gov/dp_web/documents/chilesrpt.pdf (accessed March 14, 2001)

U.S. Department of Energy, Office of Defense Programs. 1995. *The stockpile stewardship and management program: Maintaining confidence in the safety and reliability of the enduring U.S. nuclear weapon stockpile.* Washington, D.C: U.S. Department of Energy, available at stsfac.mit.edu/projects/sbss/doe.html (accessed March 14, 1999).

———. National Nuclear Security Administration, Livermore Site Office. *Fiscal year 2004, Annual performance evaluation and appraisal for Lawrence Livermore National Laboratories.* Oakland, Calif.: Livermore Site Office.

———. National Nuclear Security Administration. 2006. *NNSA's reliable replacement warhead program will enable transformation of the nuclear weapons complex*, available at www.nnsa.doe.gov/Transforming_the_Stockpile.pdf#search=%22NNSA%E2%80%99s%20reliable%20replacement%20warhead%20program%20will%20enable%20transformation%20of%20the%20nuclear%20weapons%20complex%22 (accessed September 4, 2006).

U.S. Department of Defense. 1995. *Annual defense report*. Washington, D.C., available at www.dtic.mil/execsec/adr95/toc.html (accessed March 16, 2000).

U.S. Senate. 1997. *The future of nuclear deterrence*. Committee on Governmental Affairs: Subcommittee on International Security, Proliferation and Federal Services. Washington, D.C.: United States Government Printing Office.

Vasquez, Carlos. 1997. "Impact Los Alamos:" Traditional New Mexico in a high tech world: Overview of project and symposia. *New Mexico Historical Review* 72: 3–14.

Weick, Karl. 1995. *Sensemaking in organizations*. Thousand Oaks, Calif.: Sage.

Weldes, J., M. Laffey, H. Gusterson, and R. Duvall. 1999. *Cultures of insecurity: States, communities and the production of danger.* Minneapolis: University of Minnesota.

Wenger, Etienne. 1998. *Communities of practice: Learning, meaning and identity*. Cambridge: Cambridge University Press.

White, Paul. 1987a. *Test bans, legislation and arms control*. Los Alamos, N.M.: Center for National Security Studies, Los Alamos National Laboratory.

———. 1987b. *United States national security policy and laboratory nuclear weapon development*. Los Alamos, N.M: Center for National Security Studies, Los Alamos National Laboratory.

———. 1988. *Observations on the nuclear test ban debate*. Los Alamos, N.M.: Center for National Security Studies, Los Alamos National Laboratory.

Wolff, Walter. 1984. *A typical Los Alamos National Laboratory underground nuclear test*. Los Alamos, N.M.: Los Alamos National Laboratory.

7

(Forever) At Work in the Fields of the Bomb[1]

Images of Long-Term Stewardship in Post–Cold War Nuclear Discourse

Bryan C. Taylor[2]

> A ship at sea may well require . . . a single captain and obedient crew. But a ship out of service, parked at the dock, needs only a caretaker.
>
> —Langdon Winner

> To be human is to care for things that don't care for you.
>
> —Edward Hoagland

U.S. citizens attempting to follow the story of nuclear weapons in the post–Cold War era have been surprised by plot twists. Despite official declaration of the "end" of the Cold War, initial speculation about a budgetary "peace dividend," and a general sigh of relief at the fading specter of superpower conflict, nuclear weapons have successfully adapted during this period. Rejecting challenges to their continued viability, U.S. officials have sustained nuclear weapons as a seemingly permanent fixture of American foreign policy and military strategy, and have commenced plans to revive their production (Pincus 2006). As a result, those weapons endure as a compelling cultural condition. Awkwardly suspended, they continue to radiate threat and paradox (Taylor 2003). However reluctantly, citizens continue to rely on discourse to interpret this condition, and to shape their participation in it.

One challenge faced by citizens in this process involves reconciling the resumption of nuclear weapons production with the ongoing remediation of dangerous contamination at weapons production sites. Here, citizens must develop a story enabling them to rationalize the simultaneity of *possessing old weapons*, *making new weapons*, and *cleaning up*. If they are to continue to enjoy citizen consent, nuclear officials must minimize potential incongruity surrounding these activities. That incongruity arises, for example, when citizens juxtapose the persistent claim that nuclear deterrence ensures "national security" with their growing awareness of the threats posed by weapons production to public health, worker safety, and the environment (Makhijani, Hu and Yih 1995). One subsequent rhetorical strategy practiced by officials, notes Masco (2004, 531), involves establishing wildlife preserves in the secure buffer zones that surround weapons production sites: "By presenting these sites as untouched in over fifty years, the [Department of Energy] seeks to redefine the value and object of that military fortification, replacing nuclear weapons systems with biodiversity as the security object of the nuclear state."

This rhetorical campaign, however, is fraught with tension. As the Department of Energy "attempts to retroactively expand its Cold War mission from nuclear deterrence to environmental protection," it obscures how radioactive fallout from nuclear weapons tests *already* pervades the global environment, and how declared boundaries between "contaminated" and "preserved" sites leak due to the mobility of animals and radionuclides (Masco 2004, 531). In this process, governmental discourse reframes nuclear risk: The enemy is no longer an external, foreign arsenal, but domestic contamination. Surveillance and control, however, remain the nuclear order of the day.

This project succeeds or fails based on the ability of associated discourse to produce effective and mutually satisfying relationships between nuclear officials and citizens. This chapter examines the role of "stewardship" discourse in shaping these relationships. That discourse is currently used by the Department of Energy (hereafter, DOE) and its stakeholders in multiple spheres of activity (Lowrie and Greenberg 2000). I focus here on the discourse of "Long Term Stewardship" (hereafter, LTS), which surrounds the development of systems at nuclear weapons production and waste storage sites that are intended to protect human health and the environment for the duration of threats posed by contamination. Viewed from a critical perspective, this discourse evokes two issues. The first issue involves the potentially incongruous articulation of "stewardship"—a term that traditionally refers to unambiguously beneficial resources—with *nuclear weapons*, which have long been feared as the ultimate technology of apocalypse (Boyer 1985; Weart 1988). This articulation may generate questions and concerns among audiences: *What does it mean to associate nuclear weapons with images of care, responsibility, and sustainability*? Second, LTS discourse invites consideration of whether and how the "lessons" of Cold War–era nuclear weapons production will be incorporated into current and future nuclear weapons production (Federation of Atomic Scientists 2002). This issue is controversial because there exist multiple and conflicting accounts of what those lessons are (or should be), including the urgent need for adequate waste storage, enhanced oversight of weapons production, and informed public debate concerning the morality and necessity of nuclear weapons for national security.

There is much at stake, then, in how nuclear stewardship discourse is produced, circulated, and taken up by stakeholders as "equipment for living" in post–Cold War culture. In this chapter, I proceed by examining historical development of the term "stewardship," and its association with nuclear weapons development. Next, I analyze key DOE and stakeholder texts to interpret and evaluate LTS discourse. I focus on how that discourse shapes relationships between the DOE and its stakeholders, thus influencing the nature and consequences of stewardship operations. Specifically, I explore an ongoing tension in this discourse between images of "stewardship" and "guardianship" that configure power relationships between these groups in different ways. I conclude by considering the relationship between LTS discourse and the commonly presumed ideals of stewardship, and by proposing further research that explores the relationship between LTS and its companion nuclear discourse of "Stockpile Stewardship." These two discourses, I argue, must be considered simultaneously if nuclear stewardship is to ethically serve the interests of dialogue and democracy.

STEWARDSHIP DISCOURSE: AN OVERVIEW

Etymologically, the word "steward" is compounded from the Old English "stig"—meaning "house"—and "weard"—meaning "warden" or "guard." Thus, a steward was originally one who guards the house. Discussion of this term in the current Oxford English Dictionary indicates that, while subsequent usage has expanded its meaning, most definitions establish that a steward does not own the resource in question, but does so on behalf of an "absent owner" (e.g., God, the King, an airline company). As such, the term depicts a delegation of authority in which an appointed figure exercises responsible care over entrusted possessions for the benefit of others.[3]

Scholarly discussions of stewardship typically invoke the "dominion covenant" that—along with other discourses such as science and capitalism—has historically shaped usage of this term in Western culture. As developed by Lynn White, Jr. (1973), this phrase describes the Judeo-Christian conception of creation in which God formed humanity in His own image, thereby giving humans authority over the natural world. Significantly, the scriptures associated with this tradition depict images of both a hierarchical "manager" and a nurturing "gardener." As such, they have been critiqued for endorsing *both* the benevolent care of fragile natural resources *and* their exploitation (Beavis 1994b; Ridd 1994; Warren 1994). Multiple images of stewardship have derived from these traditions, including the vigilant, prudent caretaker concerned with sustainability and biodiversity; the amoral functionary concerned with dutiful compliance with regulations; and the greedy developer concerned with thinly rationalized self-interest and the extraction of surplus value.

This etymology suggests three relevant conditions. First, "stewardship" is a catalytic term (and one so heavily invested with figurative meaning that we may consider it a *trope*) that has generated considerable *discourse*. As discourse, stewardship is elaborated as "a connected set of concepts, statements, terms and expressions which constitutes a way of talking and writing about a particular issue, thus framing the way people understand and act with respect to that issue" (Watson 1995, 816). Lowrie and Greenberg's (2000) survey of various stewardship discourses, for example, found recurring themes of "wisdom," "conservation," "responsibility," "commitment," "accountability," and "trustworthiness" (see also Burger and Gochfeld 2001, 440). Stewardship discourse encodes a distinct system of knowledge, and circulates widely in culture through various media, sites, and practices (e.g., organizational training sessions). In addition to *describing* phenomena, stewardship discourse also *prescribes* how cultural members may speak and think about them, specifying particular kinds of speakers and statements as credible. Stewardship discourse thus organizes power relations between its associated elements (Palmer 2003).

Second, the concept of stewardship has been appropriated by speakers in historically and culturally specific situations to serve their unique, evolving needs (Lerner 1994; Paterson 2001; Smith 1994). As a result, stewardship discourses are *multiple* and *polysemic*. They generate diverse meanings across their contexts of use. The relationship between these meanings may be alternately complementary and competitive. As speakers select and configure these potential meanings, stewardship becomes a "dance between the structuring properties of . . . rhetorical forms . . . and the innovative agency of practical political actors using them" (Williams 1999, 4).

Finally, stewardship discourses are *mythical narratives*: They are animated by potent cultural meanings associated with controversial matters of history, progress, nature, justice, production, consumption, governance, and participation (Harrison and Burgess 1994; Hutchinson 1994). As such, these narratives typically perform three functions (White 1980). First, they *identify* the subjects and objects associated with a particular system of "stewardship." Second, they *establish the relationships* that exist between these elements (by identifying, for example, to whom a steward pledges loyalty). Finally, they *morally evaluate* those relationships (e.g., as one of "fealty" between a steward and the owner of a resource). In this process, stewardship discourses encode cultural beliefs, values, and "lessons" concerning these elements and their relationships. For example, these discourses may depict nature as essentially malleable and beneficially engineered by technocrats. Alternately, they may ensure that stewards remain "vulnerable" to oversight, dialogue, and reflection (Davis, Schoorman and Donaldson 1997, 42). In this process, stewardship discourses articulate "artifacts" with "politics" (Winner 1980, cit. Feaver 1992, 23–24).[4] They reproduce and transform the ideological narratives of their associated institutions—for example, by depicting those institutions as altruistic, inclusive, skillful, egalitarian, collaborative, and flexible (as opposed to exclusive, incompetent, hierarchical, greedy, unilateral, and rigid). As a result, stewardship discourses may become a site of struggle between institutions and stakeholder groups, as their members utilize preferred dialects to assert and advance their respective interests (Peterson and Horton 1995).

STEWARDSHIP DISCOURSE IN NUCLEAR CULTURE

Stewardship discourse has been uniquely inflected by nuclear history. That history involves the development of complex institutions and technologies to manage the inter-related phenomena of nuclear weapons, power, and waste. Because these phenomena possess dangerous material and symbolic potency (Kinsella 2001; Nadel 1995), nuclear cultures have appropriated stewardship discourses to "contain" their effects. These discourses prescribe how the elements of nuclear systems should be configured—for example, by specifying the distance from population centers that production reactors may be safely sited. The term "stewardship" appears to have first entered the official DOE lexicon in 1994, during the period of then-Secretary Hazel O'Leary's "Openness Initiative." Here, the term encoded the agency's post–Cold War sense of responsibility for cleaning up contaminated weapons production sites and restoring them for public use (Mayer and Greenberg 2002, 141). A formal stewardship program was instituted by the DOE in 1996, and was directed to achieve sustainable development at its land and facilities through ecosystem management. As envisioned, that program would preserve valuable national resources, and "integrate mission, economic, ecological, social and cultural factors in a comprehensive plan for each site that will guide land and facility decisions" (Burger and Gochfeld 2001, 441). The term was also adopted during this period to describe two other DOE programs devoted to managing the nation's nuclear arsenal ("Stockpile Stewardship"), and its inventory of fissile materials used or stored at sites ("Nuclear Materials Stewardship") (Lowrie and Greenberg 2000). As a result, "stewardship was used in a variety of ways to mean a variety of different things depending on the perspective and goals of the program actors or policy authors" (Greenberg et al. 2002, 68).

Nuclear stewardship discourses are animated by larger cultural conflicts surrounding the traditions and consequences of technological development. These conflicts often focus on the sequestered character of that development in capitalism by technocratic elites, and its devastating consequences (both actual and potential) for the welfare of humans and Nature (Caldwell 1999). Within these conflicts, nuclear weapons have acquired totemic status as "a condensed symbol for the worst of modernity" (Weart 1988, 392). They have become a token for conducting root-cultural conflicts surrounding the relationships between "authorities" and "victims" (e.g., between expertise and dissent), "logic" and "feeling" (e.g., rationality and intuition), and "nature" and "culture" (e.g., the paradigms of sustainability and development). In his ecofeminist critique of stewardship discourse, for example, Warren (1994) cites Carol Cohn's (1987) famous analysis of "sex and death in the rational world of defense intellectuals" as an example of how sexist-naturist language pervades the discourse of nuclear-professionals, and contributes to a "dysfunctional patriarchal system." By extension, we are to conclude, any stewardship discourse developed by those professionals will be ethically compromised, because it is saturated with rigid and hierarchical values.

Because it is inherently future-oriented, however, stewardship discourse can serve a valuable purpose in these debates. Potentially, it contrasts the limited temporal horizons of impulsive self-gratification and accelerated planning with farsighted prudence that is accountable to the needs of subsequent generations (Binde 2000). As represented in stewardship discourse, the nuclear Other is not an embodied difference immediately engaged in contemporary space (e.g., as one travels across international cultures). Instead, that Other is speculatively invoked as a pre-emptive, haunting voice, directed backward in time from a threatened future to motivate nuclear actors in the present.[5] Writing at a peak of renewed Cold War tensions, for example, Mormon legal scholar Edwin Firmage (1983) invoked a theology of stewardship to oppose "idolatrous" worship of the "false gods" of technology and national security that threatened human obligations to future generations.

NUCLEAR STEWARDSHIP AND GUARDIANSHIP

Here, we must briefly consider one recurring image in this history—that of nuclear "guardianship"—because it consistently shadows and potentially contradicts the benevolent face of nuclear stewardship. Historically, this image arises from a logical entailment constructed in nuclear discourse between the sublime "mystery" of nuclear phenomena and the quasi-theological authority of officials charged with their understanding and control (Kinsella 2005). More specifically, it has been developed in accounts of the tense relationship between civilian and military systems of control of U.S. nuclear weapons (Born, forthcoming; Feaver 1992; Nolan 1989).[6] Dahl (1985) argues that guardianship has been the *de facto* system of U.S. nuclear governance in the Cold War era. In this system, a minority of technocratic and military elites contrasts their expert knowledge and patriotic commitment with that of the general citizenry. Because these elites control the ideological terms on which that contrast is performed, they are able to conclude that the needs of citizens are best served by the elites' exclusive control of decisions concerning nuclear security and risk. In presuming that citizens are inherently unqualified to participate in arcane matters of nuclear governance, and in perpetuating conditions of secrecy to inhibit that participation, guardianship is both antidemocratic and autonomous (Garfield 1994). It fosters an authoritative "priesthood"[7] culture among nuclear professionals, in which their sense of entitlement and devotion buffers them from the challenges of popular oversight and dissent. In the discourse of guardianship, formal "responsibility" for and "custody" of nuclear weapons are infused with a solemn morality.

Here, we can note two conditions that shape the hegemony of nuclear guardianship. The first is that, while it pervades a broad system of nuclear institutions, guardianship is not monolithic or static. Opportunities for citizen voice vary across the nuclear landscape. Negative publicity, stakeholder pressure for reform, and Congressional legislation, for example, motivated the DOE in the early 1990s to create new structures for citizen participation in the development of site cleanup plans. The second caveat is that nuclear professionals may adopt the discourse of stewardship (i.e., of the "good servant"; Bailey 1995) *even while preserving* the institutions of guardianship. That is, within the conflicted logic of deterrence, nuclear professionals commonly view themselves as acting in a selfless manner, and laboring for the good of the nation by enhancing its security (Rosenthal 1990; Shroyer 1998). This selflessness satisfies one required element of stewardship: "To willingly act as steward for another, one must care about the person(s) whose interests are thus served" (Welchman 1999, 415). As a result, although they may logically conflict, the discourses of nuclear stewardship and guardianship are frequently articulated together.

So far, our discussion has generated two insights that we should carry forward. The first is that nuclear stewardship is not a monologic discourse that serves a single set of political interests such as environmentalism. Indeed, as a vehicle for ideology, that discourse is multi-voiced and promiscuous. The second is that a dialectical tension exists between the ideals of "stewardship" and the traditions of "guardianship," as they manifest in LTS discourse. In using these insights to examine that discourse, we may anticipate encountering four conditions of post–Cold War discourse surrounding nuclear weapons development.

The first condition involves persistent "nukespeak" that rationalizes nuclear weapons and their legacies through the discourses of "bureaucratization" and "domestication" (Schiappa 1989). A second condition involves competing discourses surrounding the risk posed by nuclear weapons development to public health and the environment. Third is a clash between the impulses of democratic openness and authoritarian secrecy surrounding oversight of nuclear weapons production (Hales 1997; Kinsella 2005). A final condition involves the ongoing narrative challenge of adequately situating nuclear weapons development in time. This challenge is produced by the radical potential of nuclear weapons to rupture history (i.e., through global nuclear war)—a scenario which grates against the ostensibly ethical orientation of stewardship to the needs of future generations (Welchman 1999, 416). In antinuclear discourse, for example, that potential is depicted as (still) threatening the emergence of those generations, and irrevocably poisoning their environment (Schell 2001). As a result, LTS discourse must situate its operations within the precarious and humbling condition of nuclear time.

Combined, these conditions suggest a conflicted environment for the development of LTS discourse. Tradition favors the discourses of guardianship seeking to maintain control over nuclear weapons development and its consequences. Those discourses have recently vied, however, with competitors valuing decentralization, collaboration, dialogue and democracy. Below, I explore how this turbulence has shaped the development and use of LTS discourse.

LONG-TERM STEWARDSHIP: AN OVERVIEW

In analyzing a discourse, it is helpful to understand how its speakers are positioned in relation to institutional structures, and how they engage each other through cultural forms and practices. As a result, we may consider the traditions, hopes, expectations, and anxieties that speakers bring to their use of that discourse. We may grasp how it expresses and configures the interests of the groups that those speakers represent. We may consider how that discourse creates, maintains and transforms their identities as group members. To these ends, I provide here a brief account of LTS discourse that describes its emergence within the DOE, its development within the contexts of controversy and litigation, and the development of specific ideals and challenges that shape its performance by post–Cold War speakers. This account is intended to provide granularity that respects the complexity of LTS discourse, and facilitates the integrity of its analysis.

Historically, "long-term stewardship" (LTS) was formalized during the mid-1990s as the DOE and its external stakeholders[8] acknowledged that planned "cleanup" activities at many contaminated sites would not be "complete." That is, even if they are successfully concluded, these activities will not allow for immediate, risk-free, or unrestricted "future land use" (associated, for example, with "cleanup-to-background" levels of radiation). Additionally, the agency acknowledged, it had no plans for managing residual hazards such as entombed waste left at some sites following cleanup.

The contributing causes of this condition were multiple and complex. Contamination at these sites is severe and widespread, and involves materials posing risks to human health and the environment for many thousands of years. Technologies currently available to remediate these sites are imperfect and evolving. Congressional funding for cleanup is limited, and competes with other national priorities. As a result, the DOE twice revised its initial definition of stewardship during this period: First, to include only limited, post-cleanup activities, and then again to accommodate both post-cleanup *and* interim activities (such as monitoring and controlling hazards) conducted at those sites where no technology or plan currently exists to complete remediation (USDOE 1999, 1, 5; 2001a, 1–2; 2001b, i). In this process, stewardship, as defined by DOE's Office of Environmental Management programs (EM) came to refer to "a stage of operations rather than an overall management orientation or culture of conducting business" (Mayer and Greenberg 2002, 141).

Simultaneously, key stakeholder groups and external research organizations invoked the term to "emphasize the need for an overall strategy to guide decisions that must be visible and understood by the public" (Greenberg et al. 2002, 68). In one usage, for example, the Nuclear Regulatory Commission emphasized the containment of hazards and reduction of risks for humans and the environment (Burger and Gochfeld 2001, 442). In another, a Hanford stakeholder group distinguished between three "levels" of stewardship: *remediation* ("to cure, relieve, or heal"); *restoration* ("to bring back to a former, original, or normal condition"), and *preservation* ("To keep safe from harm. . . . To maintain and reserve for use") (Leckband 2000). These alternate usages did not necessarily oppose DOE definitions (Burger and Gochfeld 2001, 443), although some stakeholders associated "a sacred responsibility" (Probst and McGovern 1998, ix) and "moral . . . obligations" (Oak Ridge 1998, 1) with stewardship. These associations implicitly challenged a mission-oriented instrumentalism in official stewardship discourse—specifying, for example, a "list of tasks or functions to be implemented" (Probst and McGovern 1998, ix; see also Rocky Flats 1999, 3, 9; Burger 2001, 392). Whether moral or instrumental, however, DOE and stakeholder concerns with "stewarding" stigmatized contamination differed from vernacular associations of this term with "maintaining functioning (and aesthetically pleasing) ecosystems" (Burger 2001, 392; USDOE 2001a, 1-1). Stakeholder discourse during this period (USDOE 2001c) reflected concerns that the DOE ensure comprehensive characterization of site contamination, expand the scope of LTS to include management of *all* elements of site ecosystems, ensure adequate funding, build successful relationships with affected state and local governments and Tribal Nations, and negotiate cleanup levels that would minimize—if not eliminate—the need for LTS (see for example, Oak Ridge 1998, 2). In this process, the goals and operations of LTS became contested matters.

Administratively, DOE addressed the need for LTS by establishing in 1999 an Office of Long-Term Stewardship within its EM division (which was itself established in 1989). LTS emerged during a period of controversy regarding DOE's effectiveness in conducting cleanup at contaminated sites. Stakeholders (including prominent scientific panels and think tank researchers) criticized DOE's initial performance in this arena, citing inexperience and wasted resources, lack of clear policy and leadership, inappropriate influence of local economic development goals, and insufficient evidence of progress to warrant large, continuous funding (Brinkley 2002; National Academy of Sciences 1995; Probst and Lowe 2000). Additionally, stakeholders criticized initial EM planning for site end-states as unilateral and inadequate, and demanded that DOE consider conducting long-term monitoring of sites (USDOE 1998, 6–7).

In response to this controversy, DOE developed and implemented a series of initiatives. These included: completing and revising baseline estimates of site cleanup costs; tightening internal contract-management practices; shifting funding from passive site surveillance and maintenance activities to actual contaminant removal and waste processing; negotiating "accelerated cleanup" agreements at some sites that specified completion dates and end-state radiation levels; and generally, seeking "to provide for smooth transition from cleanup to long-term stewardship through technical, financial, and managerial planning" (USDOE 1999, 2; see also 2001b, 106). These initiatives were partly spurred by DOE's December 1998 settlement of a lawsuit brought by 39 groups against the agency's decision to exclude planning for environmental restoration activities from its cleanup program. This settlement agreement required the agency to prepare an Internet database of its waste holdings, to fund citizen monitoring and independent technical assessments of its cleanup activities, and to prepare a comprehensive study for Congress of LTS requirements. While surveying existing LTS plans at its sites, DOE held scoping workshops in 1999 to involve stakeholders in the preparation of this study. Stakeholder groups—including the DOE-EM Site Specific Advisory Boards (hereafter, SSABs)—met both independently and jointly during this period to develop their positions on LTS. DOE claims that throughout this scoping process, it "made a concerted effort to address every substantive comment received," save for site-specific issues not suitable for the preparation of broad policy (USDOE 2001b, B–2). In addition to drafting (USDOE 2001a) and revising (USDOE 2001b) that study, the agency formed an Executive Steering Committee and Working Group within its Office of LTS to facilitate strategic planning. One initial task for that group involved "developing a common, consistent understanding of the definition and scope of long-term stewardship" (USDOE 2001b, 36). Structural and cultural changes followed as these agents assigned responsibility for LTS to specific DOE units at each site, prepared guidance for sites to develop local LTS plans, revised the agency's larger planning and budgeting systems, developed directives to ensure agency staff and contractor compliance with LTS requirements, provided associated training, and developed mechanisms for cultivating public involvement. DOE programs for LTS at completed-cleanup sites, and for associated worker and community transition were relocated in the spring of 2003 to a new unit, the Office of Legacy Management.

While many stakeholders welcomed these developments, others criticized them as encouraging compromised cleanups in which higher levels of residual risk are traded to secure a brief, steady flow—but reduced total amount—of federal funding (Dayton 2002; Moore 2005; "Shortcut on Nuclear Waste" 2004). Since 1989, DOE's annual funding for environmental cleanup and waste management programs has risen from $1.6 to $6 billion; that funding is projected to total around $300 billion at the end of their lifecycle in 2070 (Mayer and Greenberg 2002). LTS costs will accrue both during and after this period as sites shift from one stage of operations to the other. Since 2000, total annual costs for LTS activities have fluctuated between $50 and $60 million (Lambrecht 2003), and are expected to stabilize after 2050 at a level between $100 and 110 million (Wells and Spitz 2003; USDOE 2001b, 3–20, 94). While the annual costs of LTS are only one-tenth of those for environmental remediation, their lifecycle is vastly longer (Mayer and Greenberg 2002, 141; USDOE 2001a, 3–19). Their total lifecycle costs are thus unknown.

During this period, DOE officials and stakeholders proposed several ideals for the process and outcomes of LTS discourse (Applegate and Dycus 1998; Drew and Nyerges 2004; Greenberg and Lowrie 2001; Mayer and Greenberg 2002; National Research Council 2000; The Perspectives Group 2002; USDOE 2001a). Speakers subsequently drew upon these ideals in evaluating the effectiveness and appropriateness of LTS discourse. Following are six ideals prominent in that discussion.

First, *officials should involve stakeholders in decision-making by providing them with sufficient opportunities for "one-way" and "two-way" communication* (USDOE 2001b, 109). Here, officials seek to ensure "broad . . . public discussion" (Rocky Flats Stewardship 1999, 8); "meaningful participation" (USDOE 2001a, 4–4); and "adequate," "continuous" (Oak Ridge 1998, 9), "early, informed, and regular" public involvement ("EMSSAB Recommendations" 2000). In this process, officials also seek to avoid creating "participation fatigue" and burnout. Here, stakeholders have requested that the DOE support "[the use of] independent technical expertise, retention of the SSAB's during long-term stewardship,[9] funding of a system of community boards, funding for active and ongoing public education, funding for community-led studies, and training members of the public to find and interpret monitoring data" (USDOE 2001b, 20), as well as the development of museums, visitors' centers, and educational facilities (The Perspectives Group 2002; USDOE 2001b, 86–87).

As a second ideal for LTS discourse, officials and stakeholders agreed that *DOE should plan for both "routine interactions with stakeholders"* (Greenberg and Lowrie 2001, 130) *and unexpected opportunities to strengthen relationships*. Third, *DOE should ensure maximum "transparency" of decision-making*. This standard should be applied to the design and administration of both *decision-making processes* and *information systems*. Those processes should be characterized by inclusive and robust stakeholder participation, full and open (e.g., "candid") evaluation of risks (e.g., including issues of intergenerational equity; Ahearne 2000), and accountability by decision makers. Information systems, in turn, will ideally display a number of characteristics, including: adequacy for the preservation of both institutional and cultural memory, currency, durability, redundancy, credibility, clarity, accessibility, accuracy, and relevance for local community needs. Those systems should also strike an appropriate balance between the needs for both flexibility and reliability on the one hand, and centralization and decentralization on the other.

As a fourth ideal, *DOE and stakeholder organizations should develop a "culture of stewardship."* Ideally, this culture would manifest as those organizations developed explicit LTS-related elements in their mission and vision statements, secured adequate funding for LTS operations, established formal visibility and authority within their institutional environments,[10] designed measurable objectives, integrated related LTS values into planning and operating new facilities, and continuously reflected on "lessons learned."

Fifth, *DOE employees should coordinate effectively to ensure that headquarters staff adequately guide the development of site programs, and ensure their consistency. Site employees, meanwhile, should ensure that program flexibility is tailored to the needs of local stakeholders.*

Finally, *organizations should ensure the sustainability of LTS by informing successive generations of stakeholders about site history and risks.* Related education and training programs should secure stakeholders' ongoing commitment to activities such as monitoring, maintenance, responding to the further discovery of hazards, and integrating future science and technology into the cleanup of residual hazards (USDOE 2001b, 119–127).

Clearly, these ideals for LTS discourse were ambitious. They could not be achieved, however, without sweeping—and potentially traumatic—organizational change. As a result, these same commentators noted nine challenges faced by LTS officials and stakeholders in their efforts to achieve these ideals. These challenges typically arose from the existence of multiple, simultaneous and competing influences on the production and reception of LTS discourse.

The first challenge involved the efforts of speakers to *distinguish LTS from other related terms and concepts*, including:

- "*Cleanup*": "The process of addressing contaminated land, facilities, and materials in accordance with applicable requirements" (USDOE 2001a, 1-1; 2001b, 13–27);
- "*End state*": "The physical condition reached when cleanup actions are complete"; (USDOE 2001b, 13, n.7)
- "*Remediation*": A synonym for cleanup, typically emphasizing activities conducted onsite (such as dismantling building process lines) as opposed to off-site (such as long-term geologic storage of waste);
- "*Institutional control*": A term traditionally used by the Environmental Protection Agency (EPA) to encompass LTS-related activities of controlling and cleaning up hazardous waste releases, performing waste stabilization and containment, monitoring system performance, maintaining engineered controls, and controlling site access through markers, land-use restrictions, public archives, and "other methods of preserving knowledge" (USDOE 1999, 15); and
- "*Future land use*": Negotiated agreements that specify the desired nature and level of human activities on a site following cleanup.

This task of distinguishing LTS was further complicated because the definitions for many of these alternate terms are *themselves* contested. Commentators also noted that these alternate terms may be organized differently as components of discourses that express competing approaches to cleanup and stewardship (Burger 2000; Burger and Gochfeld, 2001; Crowley and Ahearne, 2002). This first challenge was closely tied to a second challenge involving *the coexistence of competing definitions within and across sites for the objects and processes of LTS activities* (USDOE 2001a, 2–17).

A third challenge to achieving ideal LTS discourse involved *the traditional reluctance of officials to explicate the values, goals and criteria that they actually use as the bases for their decisions*. This reluctance potentially affected stakeholder perceptions of the logic, accuracy, openness and accountability of LTS deliberations (National Research Council 2000, 95). A fourth challenge arose from the condition that *inevitable trade-offs exist between different values associated with the ideal of transparency* (such as openness and privacy), *and also between that ideal and the outcomes of its assertion in decision-making*. For example, insistence on transparent deliberations may lead to increased posturing in public forums by elected officials, and the refinement of political tactics used by speakers to conceal information and frustrate audience understanding. A fifth challenge involved *the sparse, obscure, fragmented, ambiguous, and technically complex qualities of LTS-related information*. Revealingly, one DOE official warned that the "mind-numbing banality" of LTS discourse inhibited public understanding (Werner 2000.[11])

A sixth challenge to achieving ideal LTS discourse arose from *dysfunctional legacies of DOE's Cold War–era organizational culture that pervaded cleanup operations*. These legacies included obsessive secrecy and security, inefficiency, fragmentation, inertia, turf-battling, "inscrutable budget allocation and accounting processes, and a lack of effective mechanisms for [ensuring] internal and external accountability" (Probst and Lowe 2000, 8). Seventh, *lingering and corrosive differences between DOE officials and stakeholder groups inhibited their development of trust and coordination*. Resistance to the LTS mission among DOE's Cold Warriors, for example, was fueled by the fact that the figure hired to direct its associated office had served as a staff member for one of the environmental groups party to the successful 1998 lawsuit against the DOE. Eighth, *stakeholder groups displayed varying communication preferences*. For example, some groups preferred to engage in passive monitoring of decision-making opportunities, while others preferred to actively participate.

A final challenge to achieving ideal LTS discourse involved *the chilling effect on official openness produced by DOE's response during the post–Cold War era to apparent security breaches and to the terrorist attacks of 9/11*. In 2000, for example, following repeated scandals surrounding the security of weapons design "secrets," DOE reorganized the Los Alamos, Sandia, and Lawrence Livermore Laboratories under a separate and restrictive National Nuclear Security Agency (NNSA). Following the terrorist attacks in 2001 on New York City and Washington, D.C., the DOE has aggressively reclassified "sensitive" LTS information, has restricted its Internet availability, and, according to activists, has inappropriately rejected and ignored related Freedom of Information Act requests (Costner 2002).

ANALYSIS: PRINCIPAL THEMES IN LTS DISCOURSE

The tension between these ideals and challenges ensured that LTS discourse would form a site of struggle between DOE officials and stakeholders. As a result, these groups have produced and consumed LTS discourse as a means to privilege the legitimacy and authority of their preferred narratives concerning the legacies of nuclear weapons production. I turn now to consider four themes in LTS discourse that clarify this maneuvering. The first theme involves the nature and consequences of relationships between "cleanup" and "long-term stewardship." A second theme involves the role of "domestication" as a dialect of "nukespeak" that shapes stakeholder understanding and participation in LTS programs. Third, I discuss the opportunities and dilemmas for stakeholder participation created by DOE's structuring of LTS discourse. Finally, I examine the consequences created for the credibility of LTS programs by their discursive situation in time. Throughout, I focus on ability of these themes to illuminate the ethical and political dynamics of LTS discourse.

The Relationship between "Cleanup" and "Stewardship"

In this first theme, I argue that the discursively constructed relationship between the terms *cleanup* and *long-term stewardship* is politically significant. Negotiated agreements concerning future land use, for example, attach end-state levels that shape contractor goals and operations during site cleanup. Those cleanups are thus considered complete when a variety of conditions are satisfied. These conditions include: designated facilities have been deactivated or decommissioned; hazardous materials released into the environment have been processed to meet agreed-upon goals and standards; groundwater contamination has been contained and long-term treatment or monitoring has been established; fissile nuclear materials have been stabilized, relocated and/or placed in safe long-term storage, and "legacy waste" from past production has been stored or disposed of in an approved manner (USDOE 2001b, 1-1). Although stewardship is typically framed as "what comes after" cleanup, these terms actually refer to interdependent—not discrete—activities. The enduring "success" of cleanup, for example, is dependent upon the effectiveness of subsequent LTS programs (USDOE 2001b, 1–5). Conversely, the scope of LTS as a hazard-*management* program is determined by the hazard-*elimination* goals that cleanup operations choose *not* to achieve (Crowley and Ahearne 2002, 523).

Here, one challenge for stakeholders is to avoid the seductive-but-potentially-misleading connotation of "cleanup" as the *elimination* of contamination (USDOE 2001b, 18). Here, LTS represents a dangerous opportunity for stakeholders: On the one hand, it evokes a positive image of commitment by the DOE to extended vigilance over a potentially unsuccessful cleanup. On the other, it offers a possible hedge by which the DOE might justify inadequate cleanups as a reasonable resolution of situational exigencies. Viewed from this perspective, whatever hazard is not removed, whatever risk is not reduced, simply rolls forward as a matter for LTS. Thus, the more that LTS is represented to stakeholders as a credible, effective regime that protects human health and the environment from risk, the more it potentially supports DOE's position in negotiating accelerated cleanup. That position might be expressed as: Don't worry; this cleanup is safe, and LTS will keep it that way.

Thus, it is politically significant whether the determination of site end-states (as the demarcation between cleanup and LTS operations) represents an undistorted, consensual agreement between the DOE and stakeholders. For once future-land use states and cleanup levels are negotiated and achieved, it is local communities—and not the DOE—who bear legal responsibility and financial cost should stakeholders wish to clean up sites further to "less restrictive land uses" (USDOE 2001a, 2–13). In this way, any representation of LTS as a utopian future of continuous technological innovation and institutional responsibility may encourage stakeholders to accept relatively riskier end-states than they might otherwise choose.

The Dialect of "Domestication" in LTS Discourse

In the second theme, I consider a dialect of "domestication" operating in LTS discourse surrounding the management of radioactive waste. As observers have noted (Cohn 1987; Schiappa 1989; Taylor 1997b), this dialect operates to distort the perception of risks associated with nuclear weapons development by associating them with positive and sacred images of the private sphere—particularly of "home." One example involves the common DOE designation of costs for passive site surveillance and maintenance activities as a "mortgage." Ackland (2001) notes another example surrounding the preparation of waste at the Rocky Flats facility, in which a contractor spokesperson adopted a familiar metaphor: "Our job right now is to package up this material and get it into the *moving boxes*. And then all that has to happen is that the *moving van*'s got to show up and take it" (emphasis added). This speaker's association is significant because it occurred during a period of political controversy surrounding the selection of a regional site to house this waste. As such it reflects a euphemistic discretion that functions to preserve precarious agreements negotiated between state governments, site operators, and federal regulators concerning the final disposition of hazardous waste. Thus, one critical stance toward LTS discourse involves maintaining a healthy suspicion that domestication pacifies audiences and supports official decision-preferences (e.g., by providing false reassurance concerning risk).

Russell (2000, 1–2) notes, however, that this influence may run in both directions. Specifically, he argues that a "reigning homely image" dominates conceptions of site cleanup. In this image,

> The country made a mess in conducting the nuclear-related tasks of World War II and the Cold War. Now it is time for a housekeeper (the DOE Environmental Management Program) to come in after the party, take out the garbage, straighten up the house, and put everything back to normal—to conduct the cleanup. Indeed the word "cleanup" permeates DOE literature even though in literal terms much of the "garbage" cannot be rendered harmless; there is no "away" to which to take it, and many of the sites can never be returned to uncontrolled use, much less restored to their pre-DOE conditions.

Russell opposes the impression created by this "myth" (2000, 25) that site cleanup *can* completely restore affected ecosystems, and that stakeholders *should* expend maximum resources to isolate, contain and remove threats. Alternately, he argues for a paradigm of "burden minimization." This paradigm involves comprehensive consideration and careful balancing of both legacy risks (e.g., posed to public health) and opportunity costs (e.g., manifest as worker injuries) for present and future generations that are created by planned cleanup operations. Ideally, this balancing creates "net burden reduction." In this paradigm, the "homely" stakeholder impulse to conduct aggressive cleanups that exhume, treat, transport, and store hazardous waste off-site potentially fails this test, and should be scrutinized carefully.

Alternately, Russell proposes that cleanup strategies be continually reassessed as conditions evolve in order to optimize burden minimization. Ideally, this preservation of discretion precludes scenarios of "abject" (e.g., coerced) and "malicious" (e.g., self-interested) compliance by operators, such that all stakeholder goals may be adequately considered. As a result, Russell recommends that DOE "embrace fully the perpetual mission of stewardship and [thus] raise its profile as a positive commitment to a responsible course of action, rather than treat[ing it] as a necessary but second-best default based on technological infeasibility or unacceptable current costs" (2000, 40). Russell's approach is potentially controversial, however, because it may activate lingering stakeholder mistrust of DOE commitment to LTS, and encourage stakeholders to demand immediate—and perhaps aggressive—site cleanups. That is, what Russell depicts as realistic acceptance of the inevitable imperfection of cleanup may play for some stakeholders as the advocacy of limited, compromised cleanup. Here, the issue hinges on the extent to which DOE and stakeholders are able to use LTS discourse in ways that create adequate trust, and balance the needs for both predictability and flexibility in LTS operations. The dialect of domestication in LTS discourse potentially shapes these outcomes.

Opportunities and Dilemmas for Stakeholder Participation

In this third theme, I argue that—in comparison to other DOE stewardship programs and due at least in part to formal mandates—LTS discourse is characterized by a greater opportunity for involvement in decision making by nuclear "owners" (i.e., citizens). Examples of these opportunities include "setting priorities for determining the levels and timing of cleanup, in managing for future land uses, and designing long-term stewardship plans" (Burger et al. 2001, 266). Subsequently, there exists greater—and more direct—dialogue, conflict, and accountability between officials and stakeholders in LTS decision-making. For example, even if it was initially forced by litigation, the DOE's collection and integration of public comment in preparation of its LTS Study (USDOE 2001a; 2001b; 2001c) is an impressive accomplishment that suggests the agency's potential for incorporating stakeholder participation in policy development.[12] As a result, DOE and its contractors may acquire among audiences positive images of—if not contrition—then perhaps implicit regret for their past mistakes, along with sobriety, humility, far-sightedness, resolve, (procedural) thoroughness, and steadfastness. As a result, the LTS steward appears relatively collaborative (e.g., in working with the EPA to determine respective regulatory responsibilities).

This impression should not be confused, however, with achievement of an ideal speech situation. In his analysis of public participation in cleanup decisions surrounding DOE's Rocky Flats site, for example, Moore (2005) argues that initial consensus among its stakeholders for "cleanup-to-background" was subsequently undermined by an agreement crafted without adequate public participation between DOE and federal and state regulators. As a result of public protest, the strikingly high (i.e., in the amount of residual contamination permitted) plutonium cleanup level initially set by that agreement was later reduced by 95 percent. Meanwhile, Moore alleges, DOE and the site contractor, along with members of Congress, secretly fixed key elements of an accelerated cleanup agreement for the site (e.g., by setting a completion date and imposing a limit on the sum that could be spent on closure activities). DOE subsequently commissioned a community-member focus-group to finalize this cleanup agreement, but then sprang on the group one year into its deliberations the preexistence of these constraints. Not surprisingly, the group's discussion was not able to overcome them, and the revised plutonium standards reflected a dubious compromise to complete a more thorough surface-level remediation, while permitting much higher levels to remain beneath a soil depth of three feet. This standard, Moore concludes, both deviates substantially from the community's original preference, and may expose future workers and visitors at the site's planned Fish and Wildlife Preserve to unnecessary risk.[13] This, then, is a cautionary tale about DOE's persistent strategy in ostensibly open forums of covertly controlling agendas and decision-making criteria. Should this strategy continue, the integrity of public participation in LTS will surely wither.

Nuclear Stewardship and Time

In this final theme, I consider the relationship between nuclear stewardship and time in LTS discourse. Here, I am concerned with issues such as how LTS discourse orients its audiences to the history, present, and future of nuclear weapons development, and how its stewards acknowledge potentially negative issues of decay and mortality that are raised by the duration of their mission.

I begin with a reminder that nuclear waste is marked by enduring "radiation." That radiation possesses physical properties (e.g., isotope decay) that induce social realities (e.g., perceived risk that compels stakeholders to regulate its disposition). As the DOE concedes, while the end of the Cold War has been declared, and the end of cleanup may be anticipated sometime around 2050, the temporal horizon of LTS is "virtually forever" (USDOE 2001a, 2–14). Nuclear "waste," then, is not easily drained of value or consequence. It compels disposal and vigilance, and also the development of stewardship as a story that institutionalizes that vigilance. As a result, LTS officials seem to be less plagued by tensions between stewardship and guardianship than their colleagues in other DOE programs: their mission is less ambiguously directed to the well-being of future generations. Further, these stewards concede a central rhetorical challenge faced by their enterprise: creating and maintaining across millennia communication systems that will adequately inform humans about the risks posed by radioactive waste permanently stored in geologic sites. These systems, planners acknowledge, must minimize human "interference" and survive the evolution of cultural memory and sign systems (Benford 1999; Sebeok 1986). The associated practical challenges (such as designing marker form and content) are enormous: "No existing institution," the DOE notes, "has yet acquired experience in protecting public health and the environment from hazards for such a long period of time" (USDOE 1999, 47).

Communication scholars may oppose the reduction of this complex issue by DOE officials to technical concerns of media selection and management (e.g., in the design and placement of barriers and warning signs). Those scholars may concede, however, the pragmatic wisdom in DOE's development of strategies (e.g., a "near-future time horizon") designed to minimize participants' experience in these programs of temporal vertigo. Here, nuclear stewards are encouraged to maintain their temporal balance by sticking to task and selectively responding to demands: "Focus on managing hazards for the near future (e.g., 30–50 years), rather than trying to manage hazards for centuries or millennia" (USDOE 2001b, 124). Russell (2000, 30) develops a related image of "rolling stewardship," in which "we can and should make decisions for ourselves and with a stretch of imagination for those who come immediately after us, but not 'forever'." In this image, LTS is like a baton that is to be handed off from one generation to the next (whose members will hopefully manage it to the best of their abilities). It is not a vast temporal horizon requiring a single, permanent strategy. In this view, such plans are needlessly stoic and grandiose, and may undermine the integrity of LTS. Russell's argument may underestimate, however, the inertia produced in multi-generational planning by initial timidity or compromise (e.g., leading to downward spirals of constraint). Here, officials appear—at least implicitly—to recognize the importance of inter-generational communication (e.g., programs of stakeholder recruitment and motivation) for preserving the continuity of LTS: "What is required is to create an institution that will survive for a generation—and that will take as its duty to assure that its functions will continue in like fashion for the next generation, and so on down the line" (Russell 2000, 33).

The representation of LTS in time is a politically charged process because it engages the volatile issue of *infinite cost*. Here, DOE officials anticipate potential challenges both to LTS programs, and more generally to the fiscal viability of developing nuclear weapons—in the past, in the present, and in the future. As a result, in official discourse, the LTS steward assumes a persona characterized by resolve, vigilance, and relative thrift (appropriately, since much of DOE's LTS discourse is at least partly directed at Congressional funding sources). Contrasting the relative insignificance of annual LTS costs with their substantial long-range accumulation, for example, DOE officials observe with enviable equanimity, "Any amount multiplied by perpetuity equals an infinite sum" (USDOE 2001a, 3–23). Lest the reader linger too long on the implications of this statement, its authors immediately follow with a reassurance: "The Department is, therefore, seeking to reduce these annual costs further," for example, through "investments in science and technology that could result in more reliable and less costly cleanup and long-term stewardship technologies." Related documents (USDOE 2002, i) indicate the agency's ongoing plan to "minimize [its] environmental liability for long-term stewardship"—consistent with regulation—through a variety of strategies. These strategies include limiting the potential influx of additional LTS sites, collaborating with stakeholders to develop alternative land management strategies, and "identifying alternative funding mechanisms and other potential liability-reducing strategies."

In this way, while it is by no means utopian (for example, in citing the inevitable decay of LTS controls) the rhetoric of DOE's principal report to Congress on LTS is also partly optimistic (2001a, 125–126). The conclusion of that report reassures readers that hazards will *eventually* degrade in order to permit unrestricted use, that advances in "robotics" and "science and medicine" may help to "mitigate" existing hazards, and that future limitations of land availability may lead communities to accommodate partially contaminated property. As depicted by this rhetoric, the future is a Very Long Time, in which some changes cannot be adequately anticipated or managed. But many can, enabling the achievement of "maximum possible protection" of human health and the environment. As a result, the report suggests, this portion of the nuclear future may be Not So Bad, after all.

CONCLUSION

This chapter has examined the role of "stewardship" discourse in shaping the depiction and understanding of legacies from Cold War–era nuclear weapons production. While noting the existence of multiple nuclear stewardship discourses, it has focused specifically on the case of "Long-Term Stewardship." I have examined how this discourse articulates the interests of the DOE and its stakeholders to the logic of stewardship. I conclude that it seeks to depict this steward as patient, humble, resolute, vigilant, and thrifty. The Long-Term Steward accepts his lot, even if it is not glamorous. He is Le Carré's (2001) "constant gardener" of nuclear legacies. With important exceptions, the Long-Term Steward appears to collaborate with stakeholders in developing and implementing reasonable policies for managing residual contamination at nuclear weapons production sites. These exceptions emerge as we note the power of this discourse to mobilize technology, money, labor, and consent. In those exceptions, we note ambiguity, contradiction, and distortion that form opportunities for political maneuvering and the advancement of interests.

In this maneuvering, DOE officials and stakeholders assert and reconcile competing versions of what is necessary and just in living with nuclear weapons "after" the Cold War. LTS discourse subsequently becomes a site of struggle between groups arguing over who has what responsibility, for how long, concerning the management of persistent hazards generated by the production of those weapons. In following this argument, idealistic readers may feel disillusioned that nuclear officials have appropriated the discourse of stewardship for their purposes. While I partly share this impulse, I also acknowledge the historical tradition in which political and economic interests have co-opted "stewardship" to legitimate utilitarian activities, and to preserve the status quo (Beavis 1994a, 1994b). In this tradition, powerful groups have often reduced stewardship to policy shorthand and cliché. Thus, if the DOE is guilty of incongruity in its stewardship discourse, it is not alone. Even environmentalist icons such as novelist Wallace Stegner have used this term to rationalize both limiting and facilitating human development of Nature (Ronald 1996). Thus, I do not presume a singular ideal of "stewardship" that is either upheld or degraded in LTS discourse. Instead, I am interested in how its polysemy is alternately activated and "fixed" by DOE officials and stakeholders to serve their respective interests. As I have demonstrated, this mobilization of stewardship discourse is marked by extraordinary complexity and ambiguity. As a result, we may respect communication as the ethical and deeply consequential human activity by which these conditions are, for better or worse, sufficiently resolved to enable political action.

I now conclude by sketching an agenda for future research that considers the relationships *between* discourses of nuclear stewardship. In this process, I return to a central theme raised in the introduction to this chapter: the relationship between discourses of *stewardship* and *guardianship*. As I have noted, LTS coexists with other stewardship discourses associated with DOE's weapons development programs. Principal among these is the Stockpile Stewardship Program (hereafter, SSP), which is charged with monitoring and maintaining the nuclear weapon stockpile to ensure its "safety" and "reliability" (e.g., through computer simulation of nuclear explosions; see the chapter by McNamara in this volume). Over its brief lifetime, the SSP has generated significant controversy over its necessity, scale and cost, and for its ambiguous contributions to the development of *new* nuclear weapons (Paine 2004). As a result, communication scholars should examine if and how these stewardship discourses interact to shape lines of argument in the nuclear-public sphere. Specifically, those scholars should consider how these discourses should be configured to serve the ideals of nuclear dialogue and democracy. According to those ideals, stakeholders and officials should engage in undistorted and mutually accountable discourse to express and consider relevant facts, beliefs, values, and norms. In this process, speakers can relativize the hegemony of depoliticized "facts," and surface for consideration unexamined premises that influence post–Cold War nuclear policy.

Here, scholars might begin by examining how nuclear stewardship discourses perform "boundary-work" that alternately encourages and discourages consideration of their inter-related phenomena. Kinsella (2005, 62) notes a general condition in which "official [nuclear] rhetoric, organizational structures and budgetary categories naturalize [the] distinction between 'defense' and 'non-defense' operations." However, "a more systemic view reveals that nuclear military strategy, weapons design and development, weapons production, waste disposal, environmental remediation, and civilian power applications are all deeply interconnected." Here, we may apply that systemic view to DOE's notoriously fragmented bureaucratic culture which, on first impression, "simply does not speak with one voice" (Richanbach et al. 1997, II-15), and whose compartmentalized units have traditionally operated with minimal coordination. As a result, rigid "stovepipes" of mission definition and oversight have insulated DOE programs from accountability for the consequences of their operations (e.g., through assignment of associated financial liability to other units), and have inhibited their assimilation of institutional change. Audiences of DOE discourse have thus been encouraged to follow its stockpile and long-term stewards down their respective narrow and firmly bounded paths.

Alternately, we may look for clues indicating that these discourses are *not* separate or distinct. Two examples may suffice to illustrate their connections: The first involves debate between the DOE and its stakeholders concerning DOE's obligation to clean up existing site contamination. A central clash in this debate concerns the relative risks and benefits historically accruing to communities surrounding weapons production facilities. One related position holds that the DOE "has a moral obligation to clean up all that has been contaminated" (Rocky Flats 1999, 9). Another holds that due to the economic and security benefits historically enjoyed by these communities,[14] they should be willing to assume some residual risk from an imperfect cleanup. Here, we see one potential benefit for the DOE of its continuing to assert the priority of nuclear weapons for "national security": This framing girds patriotic audiences to rationalize associated environmental costs from future weapons development programs. That sword cuts both ways, however: Recently, some residents of communities surrounding weapons production sites have attempted to influence Congressional audiences by depicting themselves as domestic-industrial "veterans" responsible for U.S. "victory" in both World War II and the Cold War (Gregoire 2002; Trever 2002). In this discourse, the sacrifices of those community members (both through their labor, and their involuntary assumption of risk) ensured national security, and justice now requires the federal government to honor that sacrifice by funding adequate cleanup. Here, as is in other scenes of controversy surrounding nuclear weapons development (Taylor 1997a; 1998), the images of honorable "veteran" and sympathetic "victim" are articulated in a mutually supportive—if precarious—relationship.[15]

A second example establishes a dialectic between SSP and LTS discourse by connecting the environmental consequences of SSP operations with the assignment of responsibility for LTS planning. Engaging this dialectic through its SSP component, antinuclear activists emphasize that SSP operations produce hazardous waste (Caldicott 2002, 43–70), and that in light of the demonstrated risk and unresolved storage problems associated with existing nuclear waste, the SSP manifests "skewed priorities" and an absurd repetition-compulsion (Slater 2001). For some of these critics, the DOE is irredeemable, and all of its operations are tainted: "The DOE's 'accelerated cleanup and long-term environmental stewardship program' has been conceived by the same agency responsible for historically placing members of the public in harm's way for national security" (Dayton 2002; see also Knutsen and Snyder 2000).

Configuring this dialectic through its LTS component, alternately, stakeholder groups have urged the DOE to ensure that LTS principles are incorporated into *all* current and future DOE operations (Citizens Advisory Board 2000; Bechtel 2000; Beachesne 2000; Marshall 2000). Here, stakeholders seek to prevent the disaster of Cold War nuclear weapons production from recurring. In this process, they recall that DOE's Defense Program (hereafter, DP) officials have in the past simply abandoned contaminated facilities, and yielded their control to EM officials. As a result, they conclude, DP officials may be unwilling or unable to integrate LTS priorities such as "pollution prevention" and planning for facility decommissioning into "national security" operations (Crandall 2000; Kelley 2000; Werner 2000).[16]

Here, stakeholders worry that the SSP drains needed cleanup funding and impedes cleanup operations at associated facilities (Community Advisory Board 2000, 17; Weida 1999, 1). They note that post–Cold War cleanup operations at some facilities may have been delayed because workers who were identified with the weapons production mission initially denied and distorted the actual extent of contamination (Pantex Plant, 2000). They scrutinize planning documents and official presentations, and conclude that—despite DOE's (2002) stated commitment to integrating LTS into all operations and appropriately limiting future LTS activities—existing incentives and accountability structures are inadequate to guarantee that future production operations will be efficient and safe (Probst and McGovern 1998, 16–20). They note with concern DOE's decree in December 2000 that responsibility for LTS following cleanup by EM officials of any future contamination at weapons production sites will be assumed by resident DP "landlords" who have historically failed to ensure the protection of health, safety, and the environment. This forced marriage between EM and DP officials is no small matter: As Greenberg et al. (2002, 68) dryly note, "It is entirely plausible that two sets of DOE decision makers with similar problems will approach and resolve them differently." DOE's recent comment on this point (2001a, 4–2) is not reassuring: "It is not yet clear whether [new weapons production] facilities are being constructed and operated in a way that will minimize or avoid the eventual long-term stewardship requirements." The likelihood that Defense Programs will not be held financially responsible for the costs of future environmental cleanup is also not encouraging here (LeRoy Moore, personal communication, 28 April, 2006). For while DOE maintains that "in cases where solutions to a problem remain elusive, prevention is the most prudent course of action" (2001a, 4–2), the agency does not consider "prevention" to include ceasing to build nuclear weapons.

Here, then, is the acid test for antinuclear and environmentalist groups: How will DOE officials express and resolve conflicts between the competing images of stewardship that are encoded in LTS and SSP discourse? How will the interests represented in these discourses be prioritized in associated planning and operations? One crucial site for that test is public involvement, where the DOE (2001c, Response Box 5.1) acknowledges that it will "balance the need to involve the public in maintaining [LTS] controls . . . with competing needs such as classified information or activities." Such cautions are not encouraging to stakeholders whose expectations for involvement were buoyed by the DOE's Openness Initiative during the 1990s. In the worst-case scenario for this trend, the LTS steward will come to resemble an earnest but compliant subordinate, engaged in the hopeless and dangerous task of managing an ever-growing volume of radioactive waste. His brethren stockpile steward, in turn, will appear hierarchically superior and resolved in generating both new weapons and waste. Here, it seems, these two sets of stewards are united in an unacknowledged complicity. Each relies on the other's "product" (waste, and a capacity for managing waste, respectively) as the resource that enables his continued operation. Thus, far from serving to inhibit future nuclear weapons production, LTS may—by appearing to successfully contain the legacies of Cold War production—foster both forgetting and repetition (Van Wyck 2001). In this scenario, LTS is normalized as a strategic "hedge" to be routinely exploited as a buffer for waste generated by future weapons production. "Interim" LTS performed at facilities where waste has not yet been stabilized or removed (e.g., Hanford and Savannah River) will continue to undermine this prospect as a testament to the extraordinary intransigence of radioactive contamination. As a result, we can anticipate that its representation may be especially contested.

This ironic condition has partly been anticipated in famous works of Cold War–era science fiction that depict post-apocalyptic societies inadvertently breaching the decaying structures of nuclear stewardship. In these narratives, members of those societies unearth and interpret textual fragments of the nuclear "secret," and repeat the history of rationalizing weapons development as the noble pursuit of "knowledge" and "innovation"—with tragic consequences (Dowling 1987). In depicting the gradual disintegration of "priesthoods" charged with guarding nuclear secrets, these novels speak to current stakeholder concerns about the relationship between SSP and LTS discourses. They question how these discourses configure—both now and in the future—the relationship between nuclear weapons development and its consequences, thereby producing extraordinary consequences for human welfare and the environment.

That these matters are not "only" fictional is established in Sebeok's (1986, 149–173) reflection on his participation in an official task force charged with designing markers for a DOE permanent waste storage facility. As a scholar aware of the inevitable decay of sign systems, Sebeok recommended that information about this site be conveyed across generations partly through a "false trail" of deliberately planted myth and superstition. Meanwhile,

> the actual "truth" would be entrusted exclusively to an—as it were— "atomic priesthood," that is, a commission of knowledgeable physicists, experts in radiation sickness, anthropologists, linguists, psychologists, semioticians, and whatever additional administrative expertise might be called for now and in the future. Membership in this elite "priesthood" would be self-selective over time (1986, 168)

The politics of Sebeok's vision lie in its fantasy concerning "self-selection" to this role. History suggests that inclusion in nuclear stewardship groups is not achieved on a level playing field.[17] As a result, Sebeok's account demonstrates that even well-intended visions of nuclear stewardship—whether of weapons or waste—can slide easily into guardianship. They do so through official presumption that nuclear "secrets" are the legitimate property of elites, who retain the right to engineer associated public understanding and involvement in decision-making.

Sebeok's narrative suggests how difficult it is for nuclear officials to assimilate cautionary tales about stewardship. To those officials, the urgency of protecting human health and the environment from further risk makes it "obvious" that they should continue to exercise their formal authority to design and manage stewardship systems. These duties are, after all, the responsibilities of their office. We should remember, however, that legal authority for nuclear stewardship is not inherently incompatible with a well-informed, motivated, and active citizenry. It is a truism of participation regimes, for instance, that increasing stakeholder involvement in decision-making increases the likelihood of achieving both accurate decisions and their successful implementation. This truism presumes, of course, that nuclear officials care whether or not citizens genuinely understand and consent to their decisions.

Applying this principle to the development of nuclear stewardship should lead us to encourage robust debate among DOE officials and stakeholders concerning the SSP and LTS programs. It should also lead us to urge that public debate about these programs *not* be conducted separately. Instead, if American citizens are to be fully informed about the past, present, and future of nuclear weapons development, they need to consider the implications of both programs simultaneously. As noted above, DOE officials have a strong incentive to account for their competing discourses of stewardship, if for no other reason than to avoid practical problems. If the stewards of these programs are not willing to manage that relationship explicitly, however, it is the responsibility of stakeholders to intervene and clarify these connections—even when they are inconvenient and discomforting.[18]

In contributing to this debate, communication scholars may accept the DOE's challenge (USDOE 1995, 90) to "look to the future and anticipate" important questions that the agency and its stakeholders must consider. As a result, we may all better understand who is serving whom in the production of nuclear weapons, and in the management of its legacies. Additionally, rather than prematurely resolving the tension between stewardship and guardianship, we may maintain a productive dialogue concerning appropriate relationships between their potential forms. Here, we may find consolation in Rosenthal's (1991, 149) study of a related struggle by Cold War–era foreign policy elites to conceptualize just forms of nuclear government:

> The realists never resolved the underlying tension that dominated their thinking about democracy and guardianship . . . Democracy and guardianship complemented each other, completing a circle: the people needed guidance from experts; the guardians needed the common sense criticism, direction, and moral consensus of the people. Neither the people nor the leaders could survive on their own.

NOTES

1. This title nods to Robert Del Tredici's 1987 groundbreaking volume of photographs documenting the U.S. nuclear weapons production complex: *At Work in the Fields of the Bomb* (New York: Harper & Row). More recently, Del Tredici's work leavens the dry text of DOE's official reports on LTS (USDOE 2001a, 2001b, 2001c). See Taylor (2003).

2. I wish to thank several individuals who read and commented on drafts of this essay, including: Len Ackland, Robert Cox, Stephen Hartnett, Judith Hendry, Lisa Keranen, Bill Kinsella, the late James McDaniel, Maribeth Metzler, Leroy Moore, Phaedra Pezzullo, and Robert Unger.

3. I thank Judith Hendry for providing these foundational definitions.

4. Space does not permit full consideration of Winner's (1980) argument that nuclear weapons exemplify how a technology's essence may compel the development of associated social and political structures. For Winner, the risks associated with nuclear weapons require the development of rigid, centralized, and authoritarian institutions of governance, and render the democratic hopes of antinuclear activists "dead wrong" (135). Indeed, Winner notes, democratic nuclear states must labor to ensure that the authoritarianism of these institutions does not "spin off or spill over into the polity as a whole" (131). Nonetheless, there are at least two reasons we need not accept this argument completely. The first is the apparent hyperbole of Winner's rhetoric, in which his pessimism seems to express both a) a general belief that technology should not be developed without consideration of the consequences for democratic life created by the development of institutions compelled by its essence, and b) a specific belief that any further development of nuclear technology subverts democracy. Here, Winner's condemnation hails its audience to develop—if not ideally democratic structures—then ones more representative, progressive and humane, thus leading to a minimization of nuclear authoritarianism. The second reason is that, in conceptualizing the landscape of nuclear technology, Winner has not adequately represented its complexity, or anticipated the consequences of DOE's legitimation crisis during the 1980s and 1990s. As noted below, the nuclear weapons production complex is vast, and the national security imperative that Winner correctly identifies as dominant is not evenly distributed or consistently influential. This volume focuses on an institutional scene in that apparatus in which national security and democratic imperatives have increasingly clashed. This chapter focuses on how discourses of stewardship alternately invite and constrain alternative politics that mediate citizen interests and nuclear risk. Its conclusion demonstrates that the intractability of guardianship haunts—but does not fully preclude—the democratic potential of nuclear stewardship discourse.

5. While this image is typically invoked to discourage present generations from offloading the costs of technological development onto the future, Russell (2000) cautions that intergenerational equity cuts both ways: Since future generations will arguably benefit from freedom and security created by nuclear weapons, it is appropriate for them to share in paying for the associated costs.

6. To complicate matters, during the 1990s an international network of environmentalist and antinuclear groups proposed "Nuclear Guardianship" as a collective-democratic system for storing and monitoring existing radioactive waste—without adding to it through further production of either nuclear weapons or power (see www.ratical.com/radiation/NGP). As proposed, this system would ensure the temporary containment of that waste, its retrieveability in the event that future generations develop superior remediation technologies, and its public visibility as a cautionary reminder of technological hubris. This history demonstrates that "guardianship" is also a polysemic term; its inflection by this network is affiliated with—not opposed to—the progressive-environmentalist traditions of "stewardship." That said, for the purposes of this study, I use the term "guardianship" to connote its authoritarian tradition that opposes progressive stewardship ideals. I thank LeRoy Moore (personal communication, 28 April, 2006) for clarifying this history.

7. This association is fostered by the fact that the longest-lived institutions available as models for LTS officials are organized religions such as the Catholic Church.

8. Including "the public and organized public interest groups, the DOE's contractors and consultants; and elected federal, state, and local government officials, and their staff, especially those involved with permitting and monitoring responsibilities" (Greenberg et al. 2002, 65). Greenberg et al., note that, while invoking hierarchical distinctions between stakeholder groups is "imprudent" (69), these groups have a contractual-legal relationship with DOE that distinguishes them from other candidates such as "residents, the media, and not-for-profit organizations."

9. This has not proven the case. As accelerated cleanup is completed at sites such as Rocky Flats, DOE has chosen to "sunset" their associated SSABs and replace them with Local Stakeholder Organizations (LSOs). It is not yet clear how the composition and administration of these organizations will affect stakeholder representation and the integrity of deliberation. See also note 14.

10. Stakeholders visiting DOE's Director of Long-Term Stewardship during this period were not reassured to find that his office in the agency's Headquarters Building was located in a small, converted closet. Two other representative anecdotes include: 1) An organizational chart submitted with a reorganization of the DOE proposed by the President's Foreign Intelligence Advisory Board in 1999 did not acknowledge the existence of Environmental Management units (Probst and Lowe 2000, 22); and 2) In 1995, DOE defunded a $5 million program for scholarships to study Environmental Management that it had established in 1992. At the same time it created a new $20 million fund for scholarships to study nuclear weapons design (Werner 2000).

11. Werner was recalling a comment made to him by the noted futurist Stewart Brand.

12. Public comments received were summarized in the Final Study (USDOE 2001b), and reproduced *in toto*—with responses—in a separate volume (USDOE 2001c). While these responses may not be satisfactory to all stakeholders, they are at least thorough.

13. More than 1,000 acres of the 6,500-acre Rocky Flats site (primarily associated with its former Industrial Zone) will be retained by the DOE, and monitored for residual contamination.

14. This view holds that local communities shared in the benefits of deterrence accruing from facility operations. Speakers making this counter-argument rarely acknowledge that such facilities were redundantly targeted for ICBM strikes by Soviet military strategists. As such, proximity was at best a mixed blessing.

15. This discussion emphasizes the progressive potential of this discourse to encourage adequate government protection of public health. Images of the Cold War "veteran" and "hero," however, are frequently invoked in triumphalist discourse to marginalize competing interpretations of Cold War history—and thus nuclear legacies (see the chapter by Krupar and Depoe in this volume).

16. Werner (2000) has characterized the resistance of DOE Cold Warriors to LTS by referencing a slogan from the progun movement: "We'll begin LTS when you pry my cold, dead fingers off our old mission"—presumably referring to weapons production.

17. At the Rocky Flats site, for example, some stakeholder groups challenged the presumption expressed by a dominant coalition of local governments that the representation of elected officials will continue be privileged in the formation of a long-term stewardship council.

18. See, for example, Len Ackland's essay "Rocky Flats II" in the September 14, 2003, Denver Post, p. E1.

BIBLIOGRAPHY

Ackland, Len. 2001. Rocky Flats: Closing in on closure. *Bulletin of the Atomic Scientists*, November/December: 52–56.

Ahearne, John F. 2000. Intergenerational issues regarding nuclear power, nuclear waste, and nuclear weapons. *Risk Analysis* 20: 763–770.

Applegate, John S., and Stephen Dycus. 1998. Institutional controls or emperor's clothes? Long-term stewardship of the nuclear weapons complex. *The Environmental Law Reporter*, 28 (November): 31–52.

Bailey, Janet. 1995. *The good servant: Making peace with the bomb at Los Alamos*. New York: Simon and Schuster.

Beachesne, Ann. 2000, 14 December. National Governors' Association comments on the Department of Energy's Long-Term Stewardship Draft Study. Comment #19 in *Long-Term Stewardship Study: Volume II—Response to public comments*, U.S. Department of Energy, 2001. Washington, D.C.

Beavis, Mary Ann. 1994a, 11–12, March. Introduction. Pp. 1–2 in *Environmental stewardship: History, theory and practice–Workshop Proceedings*, ed. Mary Ann Beavis. Winnipeg, Manitoba: University of Winnipeg Institute of Urban Studies.

———. 1994b, 11–12 March. Environmental stewardship in history, theory and practice. Pp. 3–8 in *Environmental stewardship: History, theory and practice–Workshop Proceedings*, ed. Mary Ann Beavis. Winnipeg, Manitoba: University of Winnipeg Institute of Urban Studies.

Bechtel, Dennis. 2000, 14 December. Comments by Clark County, Nevada, Department of Comprehensive Planning, Nuclear Waste Division on the Department of Energy's Long-Term Stewardship Draft Study. Comment #23 in *Long-Term Stewardship Study: Volume II—Response to public comments*, U.S. Department of Energy, 2001. Washington, D.C.

Benford, Gregory. 1999. Part One: Ten thousand years of solitude, in *Deep time: How humanity communicates across millennia*. New York: Avon Books.

Bindé, Jerome. 2000. Toward an ethics of the future. *Public Culture* 12: 51–72.

Born, Hans. (Forthcoming). Civilian control and democratic accountability of nuclear weapons. Chap. 6 in *Governing nuclear weapons: Democratic accountability and civilian control of nuclear weapons*, ed. Hans Born and Heiner Hanggi. Geneva: Geneva Centre for the Democratic Control of Armed Forces.

Boyer, Paul. 1985. *By the bomb's early light: American thought and culture at the dawn of the atomic age*. New York: Pantheon Books.

Brinkley, Joel. 2002. Energy Dept. contractors due for more scrutiny. *New York Times,* 24 November: Sec. 1, p. 28.

Burger, Joanna. 2000. Integrating environmental restoration and ecological restoration: Long-term stewardship at the Department of Energy. *Environmental Management* 26: 469–478.

———. 2001. Stewardship and future land use at a Department of Energy site: Does self-interest determine ratings? *Journal of Toxicology and Environmental Health, Part A* 63: 383–395.

Burger, Joanna, and Michael Gochfeld. 2001. Stewardship and the US Department of Energy: Encompassing ecosystem protection. *Journal of Environmental Planning and Management* 44: 437–454.

Burger, J., J. Sanchez, D. Roush, and M. Gochfeld. 2001. Risk perception, future land use and stewardship: Comparison of attitudes about Hanford Site and Idaho National Engineering and Environmental Laboratory. *Journal of Environmental Management* 61: 265–280.

Caldicott, Helen. 2002. *The new nuclear danger: George W. Bush's military-industrial complex*. New York: New Press.

Caldwell, Lynton K. 1999. Is humanity destined to self-destruct? *Politics and the Life Sciences* 18: 3–14.

Citizens Advisory Board, Idaho National Engineering and Environmental Laboratory. 2000, 15 November. Comments on the Department of Energy's Long-Term Stewardship Draft Study. Comment #7 in *Long-Term Stewardship Study: Volume II—Response to public comments*, U.S. Department of Energy, 2001. Washington, D.C.

Cohn, Carol. 1987. Sex and death in the rational world of defense intellectuals. *Signs* 12: 687–718.

Community Advisory Board for Nevada Test Site Environmental Management Program. 2000, October 25–28. Stewardship briefing book for Nevada sites. Prepared for the U.S. Department of Energy Site-Specific Advisory Board Stewardship Workshop, Denver, Colorado.

Costner, Brian. 2002. Access denied. *Bulletin of the Atomic Scientists*, March/April: pp. 58–62.

Crandall, Kathy. 2000, 30 November. Alliance for Nuclear Accountability comments to the Department of Energy's Long-Term Stewardship Draft Study. Comment #5 in *Long-Term Stewardship Study: Volume II—Response to public comments*, U.S. Department of Energy, 2001. Washington, D.C.

Crowley, Kevin D, and John F. Ahearne. 2002, November–December. Managing the environmental legacy of U.S. nuclear weapons production. *American Scientist* 90: 514–523.

Dahl, Robert. 1985. *Controlling nuclear weapons: Democracy versus guaradianship*. Syracuse: Syracuse University Press.

Davis, James H., F. David Schoorman, and Lex Donaldson. 1997. Toward a stewardship theory of management. *Academy of Management Review* 22: 20–47.

Dayton, Susan. 2002. Statement of Susan Dayton, Director, Citizen Action / New Mexico. In Appendix II: Additional material submitted for the record. U.S.

Senate Committee on Energy and Natural Resources Hearing, 107th Cong., 2d sess. 11 July.

Dowling, David. 1987. *Fictions of nuclear disaster*. Iowa City: University of Iowa Press.

Drew, Christina H., and Timothy L. Nyerges. 2004. Transparency of environmental decision making: A case study of soil cleanup inside the Hanford 100 area. *Journal of Risk Research* 7: 33–71.

Environmental Management Site-Specific Advisory Boards recommendations on long-term stewardship. 2000, October 31. Copy in possession of author.

Feaver, Peter. 1992. *Guarding the guardians: Civilian control of nuclear weapons in the United States*. Ithaca, N.Y.: Cornell University Press.

Federation of Atomic Scientists. 2002. Expectations for the U.S. nuclear stockpile stewardship program, available at www.fas.org/nuke/control/ctbt/text/foster01.doc (Accessed May 23, 2003).

Firmage, Edwin B. 1983. Allegiance and stewardship: Holy war, just war, and the Mormon tradition in the nuclear age. *Dialogue*, 16: 47–61.

Garfield, Susan. 1994. "Atomic priesthood" is not nuclear guardianship. *Nuclear Guardianship Forum*, Spring, Issue # 3, available at www.ratical.com/radiation/NGP/AtomPriesthd.html (Accessed October 18, 2001)

Gregoire, Christine. Prepared statement of Christine Gregoire, Attorney General, State of Washington. In Appendix II: Additional material submitted for the record. U.S. Senate Committee on Energy and Natural Resources Hearing, 107th Cong., 2d sess. 11 July.

Greenberg, Michael, and Karen Lowrie. 2001. A proposed model for community participation and risk communication for a DOE-led stewardship program. *Federal Facilities Environmental Journal*, Spring: 125–141.

Greenberg, M., K. Lowrie, H. Mayer and M. Frisch. 2002. External stakeholders' influence on the DOE's long-term stewardship programs. *Federal Facilities Environmental Journal*, Spring: 65–75.

Hales, Peter B. 1997. *Atomic spaces: Living on the Manhattan Project*. Urbana: University of Illinois Press.

Harrison Carolyn M. and Jacquelin Burgess. 1994. Social constructions of nature: A case study of conflicts over the development of Rainham Marshes. *Transactions of the Institute of British Geographers* 19: 291–310.

Hoagland, Edward. 2002. Even the giant squid. *Harper's Magazine*, November: 15–19.

Hutchinson, Roger. 1994. Environmental stewardship: A Christian theological understanding. Pp. 9–17 in *Environmental stewardship: History, theory and practice–Workshop Proceedings*, ed. Mary Ann Beavis. Winnipeg, Manitoba: University of Winnipeg Institute of Urban Studies.

Kelley, Marylia. 2000, 12 December. Tri-Valley CAREs comments regarding the DOD/EM/LTS Long-Term Stewardship Study Draft dated October 2000. Comment #6 in *Long-Term Stewardship Study: Volume II—Response to public comments*, U.S. Department of Energy, 2001. Washington, D.C.

Kinsella, William J. 2001. Nuclear boundaries: Material and discursive containment at the Hanford nuclear reservation. *Science as Culture* 10: 163–194.

———. 2005. One hundred years of nuclear discourse: Four master themes and their implications for environmental communication. Pp. 49–72 in *Environmental Communication Yearbook 2*, ed. Susan L. Senecah. Mahwah, N.J.: Lawrence Erlbaum Associates.

Knutsen, Reinard and Susi Snyder. 2000, 15 December. Shundahai Network's comments on the DOE's Draft Long Term Stewardship Study. Comment #30 in *Long-Term Stewardship Study: Volume II—Response to public comments*, U.S. Department of Energy, 2001. Washington, D.C.

Lambrecht, Bill. 2003. State presses U.S. on Care of Weldon Spring. *St. Louis Post Dispatch*, 16 November: A6.

Le Carré, John. 2001. *The constant gardener*. New York: Scribner.

Leckband, Susan. 2000, October 25–28. Long-Term Stewardship Update. Prepared for the U.S. Department of Energy Environmental Management Site Specific Advisory Board Long Term Stewardship Workshop, Denver, Colo..

Lerner, Sally. 1994. Local stewardship: Creating an environmental vanguard. Pp. 93–108 in *Environmental stewardship: History, theory and practice—Workshop Proceedings*, ed. Mary Ann Beavis. Winnipeg, Manitoba: University of Winnipeg Institute of Urban Studies.

Lowrie, Karen, and Michael Greenberg. 2000. The meaning of stewardship: Implications for the DOE long-term stewardship planning, Report 39. CRESP Researcher Report #39, available at www.instrm.org/cresp1/cresp1/sludge.html (Accessed May 7, 2003)

Makhijani, Arjun., Howard Hu, and Katherine Yih. 1995. *Nuclear wastelands: A global guide to nuclear weapons production and its health and environmental effects*. Cambridge, Mass.: MIT Press.

Marshall, Tom. 2000, 15 December. Rocky Mountain Peace and Justice Center's comments on the DOE's Draft Long-Term Stewardship Study. Comment #24 in *Long-Term Stewardship Study: Volume II—Response to public comments*, U.S. Department of Energy, 2001. Washington, D.C.

Masco, Joseph. 2004. Mutant ecologies: Radioactive life in post–Cold War New Mexico. *Cultural Anthropology* 19: 517–550.

Mayer, Henry J., and Michael Greenberg. 2002. Infrastructure planning in an uncertain environment: The nation's former nuclear weapons production sites. *Public Works Management & Policy* 7: 138–154.

Moore, LeRoy. 2005. Rocky Flats: The bait-and-switch cleanup. *The Bulletin of the Atomic Scientists* 61 (1): 50–57.

Nadel, Alan. 1995. *Containment culture: American narratives, postmodernism, and the atomic age.* Durham, N.C.: Duke University Press.

National Academy of Sciences. 1995. *Improving the environment: An evaluation of the DOE's Environmental Management program*, available at bob.nap.edu/html/doeemp/part1.html (Accessed April 16, 2004).

National Research Council. 2000. *Long-term institutional management of U.S. Department of Energy legacy waste sites*. Washington, D.C.: National Academy Press.

Nolan, Janne. 1989. *Guardians of the arsenal: The politics of nuclear strategy*. New York: Basic Books.

Oak Ridge Reservation End Use Working Group. 1998. *Stakeholder Report on Stewardship*, available at www.oakridge.doe.gov/em/euwg_stwd/EUGJUL98.PDF (Accessed July 20, 2002).

Paine, Christopher E. 2004. *Weaponeers of waste: A critical look at the Bush administration Energy Department's nuclear weapons complex and the first decade of science-based stockpile stewardship*, available at www.nrdc.org/nuclear/weaponeers/weaponeers.pdf (Accessed May 10, 2005).

Pantex Plant Citizens' Advisory Board. 2000, October 25–28. Program, planning and purpose for long-term environmental stewardship at Pantex. Prepared for the U.S. Department of Energy Site-Specific Advisory Board Stewardship Workshop, Denver, Colorado.

Palmer, Len. 2003. Discourses of sustainability: A Foucauldian approach, available at www.regional.org.au/au/apen/2003/2/020palmerl.htm (Accessed February 11, 2004).

Paterson, John. 2001. Institutional organization, stewardship, and religious resistance to modern agricultural trends: The Christian farmers' movement in the Netherlands and in Canada. *Agricultural History* 75: 308–328.

Perspectives Group, The. 2002, October. *Telling the story of Fernald: Community based stewardship and public access to information* (Prepared for the Fernald Citizens Advisory Board), available at www.fernaldcab.org/FRReport/FR_1.pdf (Accessed December 9, 2003).

Peterson, Tarla R., and Cristie C. Horton. 1995. Rooted in the soil: How understanding the perspectives of landowners can enhance the management of environmental disputes. *Quarterly Journal of Speech* 81: 289–308.

Pincus, Walter. 2006. U.S. plans to modernize nuclear arsenal. *Washington Post*, 4 March: A2.

Probst, Katherine N., and Adam I. Lowe. 2000. *Cleaning up the nuclear weapons complex: Does anybody care?* Washington, D.C.: Resources for the Future.

Probst, Katherine N., and Michael H. McGovern. 1998. *Long-term stewardship and the nuclear weapons complex: The challenge ahead*. Washington, D.C.: Resources for the Future.

Richanbach, P.H., D.R. Graham, J.P. Bell, and J.D. Silk. 1997. The organization and management of the nuclear weapons program (IDA-P-3306). Alexandria, Va.: Institute for Defense Analyses.

Ridd, Carl. 1994. Imagining the world, doing justice. Pp. 27–41 in *Environmental stewardship: History, theory and practice—Workshop Proceedings*, ed. Mary Ann Beavis. Winnipeg, Manitoba: University of Winnipeg Institute of Urban Studies.

Rocky Flats Stewardship Dialogue Planning Group. 1999, April. *Beyond closure: Stewardship at Rocky Flats*. Copy in possession of author.

Ronald, Ann. 1996. Stegner and stewardship. Chap. 9 in *Wallace Stegner: Man and writer*, ed. Charles E. Rankin. Albuquerque: University of New Mexico Press.

Rosenthal, Debra. 1990. *At the heart of the bomb: The dangerous allure of weapons work*. Reading, Mass.: Addison-Wesley.

Rosenthal, Joel H. 1991. *Righteous realists: Political realism, responsible power, and American culture in the nuclear age*. Baton Rouge: Louisiana State University Press.

Russell, Milton. 2000. Reducing the nuclear legacy burden: DOE environmental management strategy and implementation. Knoxville, Tenn.: Joint Institute for Energy and Environment (JIEE/2000-01), available at www.jiee.org/ (Accessed November 3, 2003).

Schell, Jonathan. 2001. *The unfinished twentieth century: The crisis of weapons of mass destruction*. New York: Verso.

Schiappa, Edward. 1989. The rhetoric of nukespeak. *Communication Monographs* 56: 253–272.

Sebeok, Thomas A. 1986. Pandora's box in aftertimes. Ch. 13 in *I think I am a verb: More contributions to the doctrine of signs*. New York: Plenum Press.

Shortcut on Nuclear Waste (Editorial). 2004. *New York Times*, 3 June: A26.

Shroyer, Jo Ann. 1998. *Secret mesa: Inside Los Alamos National Laboratory*. New York: John Wiley and Sons.

Slater, Alice. 2001, April. From the Manhattan Project to the Bronx Project: Cleaning up the toxic legacy of the nuclear age, available at www.gracelinks.org/nuke/view point/?vp=2001-04-01 (Accessed May 13, 2004).

Smith, William L. 1994. Environmental impacts, stewardship and development: The building of a religious community in Montana. *Syzygy* 3: 231–240.

Taylor, Bryan C. 1997a. Shooting downwind: Depicting the radiated body in epidemiology and documentary photography. Pp. 289–328 in *Transgressing scientific discourses: Communication and the voice of other*, ed. Michael Huspek and Gary Radford. Albany: SUNY Press.

———. 1997b. Home zero: Images of home and field in nuclear-cultural studies. *Western Journal of Communication* 61: 209–234.

———. 1998. The bodies of August: Photographic realism and controversy at the National Air and Space Museum. *Rhetoric and Public Affairs* 1: 331–361.

———. 2003. "Our bruised arms hung up as monuments": Nuclear iconography in post–Cold War culture. *Critical Studies in Media Communication* 20: 1–34.

Trever, Kathleen E. 2002. Statement of Kathleen E. Trever, Manager, Idaho INEEL Oversight Program. In Appendix II: Additional material submitted for the record. U.S. Senate Committee on Energy and Natural Resources Hearing, 107th Cong., 2d sess. 11 July.

U.S. Department of Energy. 1998. *Accelerating cleanup: Paths to closure*. Washington, D.C.

———. 1999. *From Cleanup to stewardship: A companion report to* Accelerating cleanup: Paths to closure *and background information to support the scoping process required for the 1998 PEIS Settlement Study*. Washington, D.C.

———. 2001a. *A report to Congress on long-term stewardship: Volume I—Summary report*. Washington, D.C.

———. 2001b. *Long-term stewardship study: Volume I—Report*. Washington, D.C.

———. 2001c. *Long-term stewardship study: Volume II—Response to public comments*. Washington, D.C.

———. 2002, June 21. Department of Energy's long-term stewardship strategic plan (Pre-Decisional Draft, Version 2.0), available at www.fernaldcab.org/Strategic Plan.pdf (Accessed May 17, 2005).

U.S. Department of Energy, Office of Environmental Management. 1995. *Closing the circle on the splitting of the atom: The environmental legacy of nuclear weapons production in the United States and what the Department of Energy is doing about it.* Washington, D.C.

Van Wyck, Peter C. 2001. American monument: The waste isolation pilot plant. Pp. 149–172 in *Atomic Culture: How we learned to stop worrying and love the bomb,* ed. Scott C. Zeman and Michael A. Amundson. Boulder: University Press of Colorado.

Warren, D. Michael. 1994. Indigenous knowledge systems and cross-cultural aspects of environmental stewardship. Pp. 118–130 in *Environmental stewardship: History, theory and practice–Workshop Proceedings*, ed. Mary Ann Beavis. Winnipeg, Manitoba: University of Winnipeg Institute of Urban Studies.

Watson, Tony J. 1995. Rhetoric, discourse and argument in organizational sense making: A reflexive tale. *Organizational Studies* 16: 805–821.

Weart, Spencer 1988. *Nuclear fear: A history of images*. Cambridge, Mass.: Harvard University Press.

Weida, William J. 1999. Nuclear weapons stockpile stewardship issue paper. Global Resource Action Center for the Environment, available at www.gracelinks.org/nuke/docs/Weida-SS-Issue-Paper-899.doc (Accessed June 23, 2004).

Welchman, Jennifer. 1999. The Virtues of stewardship. *Environmental Ethics* 21: 411–423.

Wells, James R., and Henry B. Spitz. 2003. Long-term stewardship of the environmental legacy at restored sites within the Department of Energy nuclear weapons complex. *Health Physics* 85: 578–584.

Werner, James. 2000, 26 October. Presentation to U.S. Department of Energy Environmental Management Site Specific Advisory Board Stewardship Workshop, Denver.

White, Hayden. 1980. The value of narrativity in the representation of reality. *Critical Inquiry* 7: 5–27.

White, Lynn, Jr. 1973. The historical roots of our ecological crisis. *Science*, 10 March: 1203–1207.

Williams, Rhys H. 1999. Visions of the good society and the religious roots of American political culture. *Sociology of Religion* 60: 1–34.

Winner, Langdon. 1980. Do artifacts have politics? *Daedalus* 109: 129–136.

Part Three

CRITICAL RESPONSE

8

Response

Nuclear Legacies and Opportunities for Politically and Ethically Engaged Communication Scholarship

Tarla Rai Peterson

This chapter proceeds from the editors' implicitly articulated assumption that public participation in democratic practice is desirable, and that ethically and politically engaged scholars have a responsibility to enhance such participation. In their introduction to this volume, for example, Taylor and Kinsella represent the U.S. Department of Energy's relative (and short-lived) communicative openness during the 1990s as "an unprecedented *opportunity* (my emphasis) for [diverse stakeholder groups] to engage each other" (22). At the same time, they note how organizational structure and political process have facilitated an arrangement between the Department of Energy (DOE) and the more recently established National Nuclear Security Administration (NNSA) that *trivializes* this opportunity by cementing the already accomplished separation between environmental remediation—a space where the public might possibly have reasonable expectations for participation, and weapons development—a space where the public must not venture, we are told, for reasons of national security. Taylor and Kinsella's basic premise is that "the nuclear-critical task is to track the structure and process" (14) of complex cultural exchanges among participants in the nuclear weapons production complex. The chapters that follow demonstrate several approaches that ethically and politically engaged scholars might use to track that structure and process.

My own research has never focused on the nuclear weapons production complex. Rather, I self-identify as an environmental communication scholar who grounds her research in practice. I view environmental communication as a problem-oriented discipline that impinges directly on the prospects for human survival, and on democracy as the suite of political practices most likely to enhance that survival. As such, I rely on critical rhetoric as a primary intellectual resource, seasoned with ideas drawn from interactions with professional colleagues and citizens of place-based communities. My research consists of subjecting intellectually compelling concepts associated with democratic practice to ground-truthing via the design, implementation, and evaluation of opportunities for public participation in environmental policy. In this chapter, I first lay out my political and ethical position regarding the relationship between democracy and communication criticism. Second, I identify and discuss potential exchanges between the chapters in this volume and attempts to conduct ethically and politically engaged scholarship in environmental communication, organizational communication, and rhetoric. Finally, I offer an admittedly personal and speculative argument for the importance of such scholarship.

DEMOCRACY AND COMMUNICATION CRITICISM

Healthy democratic process requires recognition of different beliefs, interests, and values and the recognition that open conflict over those differences is not only legitimate, but desirable (Laclau and Mouffe 2001; Mouffe 1993, 1996, 2000). From this perspective, the goal of politics in most nominal democracies, such as those found throughout Europe and the United States, should be to create a modicum of political unity through ongoing engagement in conflict, or what Laclau and Mouffe (2001) describe as a healthy dynamic between consensus and dissent. Members of various polities recognize ongoing opposition, but rather than naming their opponents as enemies, they name opponents as legitimate adversaries endowed with the same rights they expect for themselves. Ivie (2002, 2004, 2005) argues that by maintaining a productive tension between cooperation and competition, this agonistic politics can decrease the potential for transforming adversaries into enemies.

A perspective that presumes no democracy without dissent privileges communication strategies that are (at least) partial, ambiguous, and tenuous; addressed to audiences that may ignore, deliberately misconstrue, or carefully examine the message (Ivie 2004, 2005; Peterson, Peterson, and Peterson 2005). Privileging dissonance ironically promotes the democratic value of tolerance by directing attention toward the problem of how political participants "might transcend themselves sufficiently to observe their own foibles even while acting strategically toward one another—that is, how they might act with maximum consciousness by rounding out their individual perspectives through verbal sparring" (Peterson, Peterson, and Peterson 2005, 765). Thus, a political perspective grounded in argument offers a realistic means of negotiating the politics of opposing identities and interests that confront one another in policy deliberations, including, but not limited to, deliberations about the nuclear weapons complex. The whole point of democracy, therefore, is to "organize human coexistence [as] always potentially conflictual" (Mouffe 2000, 101).

This political perspective is especially important to critics of powerful formations such as the nuclear weapons production complex. Although an emphasis on civility and order, at the expense of argument, can effectively maintain existing hierarchies, it is remarkably ineffective at changing them (Mouffe 1993, 2000). Moreover, those who currently hold the reins of power rarely are willing to give it up, and typically have more patience than those seeking change, because part of their elite status comes from the fact that they have access to more resources (Ivie 2002; Mouffe 1993). Thus, Ivie (2004, 21) argues that, "democratic dissent in a period of war or crisis is as alarming to the purveyors of prevailing opinion as it is critical to a nation's political welfare." Dissent enables, although it does not ensure, resistance to dominant elites. Without it, power relationships become conflicts of interest reconcilable through mutual good will; and dogma associated with the so-called free market, the inevitability of nuclear proliferation, or unavoidable globalization, becomes reality (Peterson, Peterson, and Peterson 2005).

Like democracy, the natural world wherein we humans are organically grounded does not assume a state of perpetual stability and balance so much as a state of perpetual change and tension. We can read that tension in the resilience of organisms that evolved long before humans, and probably will observe our self-induced extinction. Rather than guard against change, they evolve ways to recover from the always present potential disaster. Terry Tempest Williams (2004) discovers in nature "a radical form of democracy at play. Each organism is rooted in its own biological niche, drawing its power from its relationship to other organisms" (47). In other words, living organisms must retain their differences, while simultaneously engaging in symbiotic relationships with others. Democracy also depends on developing, evolving, and respecting appropriate relationships across difference. Human hubris, however, mitigates against such an attitude. In a world where political leaders encourage their constituencies to invent themselves as the ultimate masters of the universe, there is no room for the amateurish practices of democracy. Democracy is messy; it makes mistakes; and its "white hot center" will never be a comfortable space (Williams 2004, 47). Attempts to make it orderly and comfortable fail, either because people resist, thus maintaining a site for struggle, or they submit, thus destroying any vestiges of democracy (Peterson et al. 2006).

CRITICISM AND ENVIRONMENTAL COMMUNICATION

Ecological problems are fundamentally material, yet at the same time their perception as problems is symbolic. Humans interact among themselves and with other species to create, compound, and (hopefully) to remediate these problems. This situation leaves environmental communication scholars with at least two related, but separate challenges. They must grapple with the dynamic nature of material/symbolic, and, in doing this, they must move beyond the human-nature dualism. In this section, I suggest how the essays in this volume might enhance environmental communication's ability to respond to these challenges, and also suggest one possibility for engagement that environmental communication suggests for nuclear criticism.

Richard Rogers' (1998) prescient essay on "transhuman materialist" rhetoric is an early response to both challenges. In addition to setting the stage for later theorizing on material rhetoric (discussed below), he suggests four criteria that could be particularly helpful to environmental communication scholars:

> (1) [T]he resurrection of a place for natural forces, traits, and structures in communication theory while avoiding a return to natural determinism; (2) an affirmation that we humans are embodied creatures embedded in a world that is not entirely [of] our own making; (3) a rehearsal of ways of listening to nondominant voices and extra-human agents and their inclusion in the production of meaning, policy and material conditions; (4) the deconstruction of common sense binaries such as subject/object, social/natural, and ideational/material, and a reconstruction of relationships as dialogic: recursive, interdependent and fluid. (268).

Nearly ten years later, communication scholars of all stripes are demonstrating increasing awareness that the interplay between material and symbolic speaks directly to their intellectual projects. Although politically engaged environmental communication scholarship should be at the forefront of theoretical investigations into this relationship, it remains peripheral to projects such as studies in material rhetoric (e.g. Blair, 1998).

Rogers (1998) also points out the folly of perpetuating a dualism between humans and nature. It attenuates efforts to avert environmental crisis by trivializing attempts to reconsider current relations between humans and Earth, including, but not limited to other animal species. Reconceptualized along the lines suggested by Rogers, environmental communication no longer would be oriented toward concerns about whether humans are unique in their ability to communicate. By debunking the myth of human exceptionalism commonly accepted in communication studies, Rogers would free environmental communication scholars to more fully explore past, present, and future relationships between humans and other creatures of the biosphere. Despite overwhelming ethical and political imperatives to do otherwise, however, environmental communication scholars have largely retained the conventional focus on humans as the communicating animal.

One reason environmental communication scholarship has failed to deeply engage the material/symbolic debate, and to transcend the human-nature dualism may be that such efforts raise questions fraught with political and ethical implications for scholars. Examples include:

- How should we articulate and evaluate environmental perturbations that emanate from human civilization's technologies, yet seem to affect extra-humans far more immediately than humans?
- How should we re-present marginalized humans and extra-humans, so as to emphasize their subjectivity, rather than their objectivity?
- How should we articulate the negotiation between material and symbolic dimensions of the biosphere and its subjects (both human and extra-human)?
- What ethical responsibilities do we have toward others, including, but not limited to, extra-humans?

Few scholars feel equipped to tackle, let alone answer, questions of this sort. However, the critiques of nuclear communication in this volume offer perspectives and tools that may enable environmental communication scholars to respond more satisfactorily to these and related questions.

Elsewhere, I have argued that Peters' (1999) preference for *dissemination* over *dialogue* as a fundamental communication mode offers a useful beginning point for environmental communication's challenge of communicating with extra-humans (Peterson et al. 2006; Peterson, Peterson, and Peterson 2007). Peters (1999) advocates replacing the ideal of dialogic communication with communication as dissemination. Although his notion of dissemination opens alarming possibilities of misdirected messages hurtling through the atmosphere, it also reframes receivers as subjects, rather than as objects to be acted upon. Thus, all beings become subjects who may (or may not) form audiences for each other (Peters, 1999). This configuration would point environmental communication toward hermeneutics, or the interpretation of texts by unintended receivers (Peterson, Peterson, and Peterson 2007). Although Peters' reframing of communication certainly is consistent with the political jumble I have argued constitutes democracy, like all perspectives, it privileges some hegemonic configurations over others. Using an internal dialogue of hermeneutics to understand disseminated representations of the environment could provide an opportunity for communication between humans and extra-human agents. On the other hand, it might mask some of the cultural formations that currently legitimize violence directed toward Earth, while at the same time labeling humans (and extra-humans) that struggle against, or simply respond to, this violence as terrorists. For example, New Orleans has been presented as the victim of natural disaster personified in Hurricane Katrina (2005). The economically and politically motivated choices to divert money intended to strengthen levies, and to destroy barrier islands that had protected the city from previous hurricanes, are rarely mentioned. Within a dissemination paradigm, the DOE is well positioned to use its website (www.energy.gov/) to tell the (potentially incongruous) story of an agency filled with capable and caring professionals to whom "the nation owes its thanks," for *both* previous weapons production and testing in the name of national security *and* heroic cleanup efforts. Those who dare question this version of events are labeled unpatriotic protesters who are guilty of venting their anger and of being irrational or, at least, unkind. That is certainly one interpretation, but it is only one.

Hamilton's essay suggests an alternative communication model that could encourage a more direct resistance to current patterns of domination, a resistance that is crucial to politically engaged environmental communication scholarship. Her analysis of a nearly twenty-year struggle between various stakeholder groups at the DOE site in Fernald, Ohio suggests the wisdom of working toward a dialogic mode, but argues that what Peters describes as the dialogic ideal is never reached, in fact, is not necessarily even sought.

The sensitivity of Hamilton's analysis is especially instructive for environmental communication scholars who seek to conduct politically relevant research within institutional and discursive systems that render some beings whole subjects, and render others (including some humans) objects. Current environmental policy responds much more positively to elite human interests than to the interests of less influential humans and extra-humans. Existing power relationships encourage objectification of the second group in service to the first. Hamilton opens a door to critiquing this process by highlighting a slow discursive shift from a culture of hostility to a dialogic culture. While the dialogic culture she describes retains space for mistrust, disagreement and anger, it also opens a learning space for legitimacy, concurrence and (always) temporary satisfaction. Her examination of how mutual discursive influences led participants to develop strong relationships that withstood frequent battering from external forces offers a model that environmental communication scholars might use in exploring possibilities for building social and political relationships that cross conventional biological, cultural, and economic barriers.

The communicative focus of several critiques collected in this volume indicates how detailed analysis of mutual discursive influence could contribute to environmental communication scholarship's ability to creatively amplify, understand, and/or resolve environmental conflict. The chapters by Hamilton, Kinsella and Mullen, and Morgan all focus on how discourse contributed to communities that formed in reaction against some perceived wrong. Hamilton's analysis illuminates a slow discursive shift from a culture of hostility to dialogic collaboration. While the dialogic culture she describes retains space for mistrust, disagreement and anger, it also opens a learning space for legitimacy, concurrence and (always) temporary satisfaction. Kinsella and Mullen examine a different sort of community, one that emerged from outrage over the realization that government officials have betrayed the public trust. Although Kinsella and Mullen's essay can be read as a heart-wrenching indictment of the DOE's blatant objectification of everything existing downwind of Hanford, it is an even more powerful testimony of the mutual discursive pain experienced as people became individual members of the downwinder community, debated over how the community would develop appropriate technical expertise, and struggled to define what that community was. Morgan's unique ethnography of communication surrounding the WIPP site suggests how environmental communication scholars might approach conflicting stakeholders as members of a community that uses dialogue to define what the local environment means as a place. All the while, they are strategically and subconsciously renegotiating that meaning in response to internal value conflicts. These three chapters offer particularly relevant contributions to politically engaged studies of environmental conflict by suggesting how mutual discursive influence contributes to a constantly shifting pattern of convergence and divergence within the evolving community, and how that pattern can facilitate publicly guided policy choices in the face of inevitable non-resolution.

Hamilton's discovery of a spatial component to the Fernald discussion, and Morgan's focus on how space becomes place are particularly relevant to two challenges that environmental communication scholars frequently face: how people conceptualize spatial scale and how space becomes place. Environmental communication scholarship has frequently engaged the concepts of space and place as a locus for political engagement and ethical commitment. The difference these essays might make is in the intellectual rigor demanded of both sets of practices. Every contributor to this volume critically evaluates processes and structures associated with the U.S. nuclear weapons complex as antidemocratic, and advocates (either implicitly or explicitly) political activism aimed at curbing its power. The authors absolutely refuse, however, to make assertions that are not empirically supported by their own studies. This approach to politically engaged scholarship might enable environmental communication scholars to avoid debacles such as the premature celebration of the Quincy Library Group (QLG), where local stakeholders formed a group that met at the Quincy Library to come up with a consensus solution to management of the forest. Scholars joined journalists in enthusiastic accolades when local representatives of the timber industry and environmental groups produced a mutually acceptable solution (Bryan and Wondolleck 2002).

When the U.S. Forest Service refused to implement the Quincy solution, it was widely vilified in the media, and scholars backpedaled furiously. Although the QLG had indeed brought together a previously divided set of stakeholders, all participants were local, and exhibited classic NIMBY (not in my backyard) characteristics. The land they proposed to manage, however, was held in trust for the entire U.S. population by the National Forest Service. Much of the QLG's initial success can be explained by the failure to include national interests in its deliberations. Much of the scholarly embarrassment can be explained by failure to unearth and critically evaluate mutual dialogic influence as a means for negotiating a productive tension between community members, as well as between the community and outsiders.

Taylor offers the same care in analyzing space, as well as another scalar issue crucial to environmental communication in his discussion of long-term stewardship (LTS). Using official DOE reports, he demonstrates how speech acts may cavalierly adjust *time* to suit political needs. Perhaps even more relevant to environmental communication scholarship is his critique of how the DOE uses the concept of stewardship (discursively articulated with time and space) to create a grandfatherly image of itself. The associated entailments for stewardship create a domestic scene wherein the DOE's intent (if not the results of its acts) must always be presumed innocent. Taylor explicitly focuses attention on processes of mutual discursive influence that lead to this presumption. Environmental communication scholars who seek to advocate responsible care of the Earth should find Taylor's treatment instructive, for the stewardship trope is everywhere used as descriptive justification for activities as diverse as housing tract development, prairie restoration, and enhanced oil recovery. Beyond an early and highly sympathetic portrayal of ranchers' self-proclaimed role as land stewards (Peterson and Horton 1995), environmental communication scholars have demonstrated surprisingly little interest in this culturally embedded and politically significant trope. Our discovery of possibilities for cooperative management with land management agencies such as the U.S. Fish and Wildlife Service retains significance today. The analysis stopped short of exploring further political significance of ranchers' deep and personal connections with their land, which overwhelmingly trumped economic motives for opposing federal control of endangered species habitat.

Environmental communication scholars need to understand the political play between stewardship, private property rights, and community responsibility. Further, perhaps the stewardship trope holds a key for transcending the entrenched human-nature dualism.

Environmental communication scholarship also offers possibilities for enhancing the political sensibilities of nuclear criticism. Cox (2007) argues that environmental communication is a crisis discipline, and offers a set of normative principles that grow out of his claim. Crisis disciplines are less inhibited (and less protected) by institutional boundaries because the urgency associated with averting impending disaster gives those disciplines license to break normative boundaries. Further, the sense of crisis promotes a deep concern with political involvement and applied ethics. Although environmental communication scholarship exhibits some of the self doubt and flagellation that haunts communication studies generally, it is not dominated by such concerns. For example, the debate showcased in the initial issues of *Environmental Communication: A Journal of Nature and Culture* (2007) focuses on how scholars can most profitably respond to environmental crisis, rather than whether they are ontologically (or epistemically, or linguistically, or . . .) *legitimated* to do so. Nuclear criticism, which I submit has all the hallmarks of a crisis discipline, could profit from a similar adjustment in debates among its practitioners.

NUCLEAR CRITICISM AND CRITICAL ORGANIZATIONAL COMMUNICATION

The nuclear criticism found in this volume also provides especially powerful lessons for critical organizational communication. Critical studies of organizational communication grew out of a neo-Marxist perspective that presumed an inherent harm to most participants in the capitalist enterprise, and an associated ethical responsibility for criticism to engage in political resistance. Contemporary scholars have struggled to defuse the predictable claims that such an adversarial position disqualifies them as "objective" researchers. One aspect of this struggle is that critical organizational communication scholars have expended a great deal of effort complicating how we understand the concepts of control and resistance. Mumby (2005) has provided a theoretical rationale for why communication scholars are particularly well positioned to provide a more nuanced explanation of how control and resistance relate to bureaucracy, domination/subordination, hierarchy, and numerous other related organizational constructs. Others have provided empirical support for Mumby's arguments. For example, Ashcraft (2005) found that airline pilots, who most people consider to hold a dominant organizational position, switch from control to resistance when the airline makes significant alterations to decision structures. She labels their decision to *consent* to organizational changes as resistant discourse. Tretheway (2005) examines how older female workers, who are generally presumed to be in less a powerful position than their younger colleagues, use narratives that potentially undermine the institutional influence of ageism.

Although these studies are valuable, particularly as they complicate organizational processes and foreground communication practices, I worry that they trivialize the researcher's ethical commitment to political engagement. The notion that control and resistance are astonishingly interchangeable is theoretically compelling. Ashcraft's characterization of pilots as resistors, rather than controllers, is interesting. Tretheway's demonstration that older females can figure out how to creatively enhance their organizational status may be heartening. When critical organizational scholars conduct research aimed at demonstrating how delightfully open organizational life is, they certainly expand their opportunities to interact with corporate decision makers. And it is likely that most researchers are much more comfortable interacting with other professionals than with day-laborers. But where are the millions of producers for this vast institutional complex? Critical organizational communication research risks becoming irrelevant to all but an increasingly small (in numbers) and elite group of humans. Beyond the potential for political triviality, I am surprised by the academic celebration of the discovery that no organizing process can entirely suture control, thus making articulation outside the current possibilities not only possible but likely. Early organizational communication studies aimed at increasing worker productivity made the same discovery, and Western Enlightenment scholars offered the same arguments.

In an essay about the nuclear weapons production complex that examines how "how planned change programs alternately promote, sustain, and defeat effective organizational change and adaptation" (390), Taylor, Kinsella, Depoe and Metzler (2005) suggest four questions that communication researchers are especially well positioned to ask. Although they are interested in these questions as they apply to nuclear weapons production, I would submit that these questions are crucial for any researchers who intend their study of organizational communication to be politically relevant beyond their own professional associations. They urge researchers to ask:

> First, what are the specific elements and processes that characterize communication surrounding planned change? . . . Second, what are the impression management strategies used by actors . . . to maintain legitimacy in a rapidly chang-

> ing institutional environment? . . . Third, how effective and ethical are these organizations in producing secondary- and meta-narration that responds to primary media coverage of crisis? . . . Finally, what are the political dimensions of organizational/stakeholder communication? (290, 291)

The critiques in this volume pursue the questions outlined above, and in so doing, suggest ways to challenge the conventional analogical claim that control is to dominance as resistance is to subordination, without cavalier dismissal of serious political realities that directly impact most people. For example, McNamara's analysis of the nuclear testing moratorium's effect on knowledge production demonstrates how research can present a nuanced picture of organizational culture, without setting aside consideration of its impact on political culture or how economic constraints play into the mix.

McNamara's explicit statement that her analysis proceeds from "a discursive regime of truth in which nuclear weapons are a beneficial proscriptive technology" signals a crucial ethical awareness on her part. To explain the almost desperate response to the moratorium among professionals at the national laboratories, McNamara turns to cultural anthropology, a relatively common resource for scholars of organizational culture. She goes beyond typical studies of organizational culture, however, in her ethnographic account of how professionals at the national laboratories have responded to the moratorium, and how this response is couched within the larger culture. The reader is treated to moments of cultural discovery, such as awareness, not only that identity is important, but of how members of the laboratories generate (and question) their identity. McNamara's ability to seamlessly integrate frequent invocation of the interrelationships between knowledge and power into descriptions of institutional culture further demonstrates the value of a traditional ethnographic approach. Too often, organizational communication scholars explain organizational culture by visiting the site every Wednesday afternoon for a semester. To further buttress the research, these organizational "ethnographies" are supplemented with interviews, questionnaires, and other attempts to discover truth through triangulation. McNamara's approach offers the reader a much richer picture of how planned change can impact identity construction, as unexpected connections that develop between political culture throughout the nation and identity construction within the organization.

McNamara's presentation of physicists, engineers, and other professionals associated with nuclear weapons production is certainly sympathetic. She offers no simplistic portrayal of dominant elites complacently protecting their turf. Like Ashcraft's pilots and Tretheway's older female professionals, their discourse is resistant. What I'm especially drawn to, however, is that, in McNamara's analysis, this is only the introduction. The big story is about how personal identity meshes with organizational interest, which morphs into a political imperative for the entire nation, in a project that has serious implications for global security.

The essays by Morgan and by Kinsella and Mullen also examine identity, but from different perspectives. Morgan's chapter focuses on how impression management strategies contribute to the assimilation of individuals into organizational projects. Like McNamara, Morgan spends time in the community, giving his conclusions about how discursive patterns develop in that community greater weight than if he had done a quick, in-and-out survey. His description of how local residents learn to talk (and think and feel) about their community and the Waste Isolation Pilot Plant as mutually constitutive is an acclamation—or a condemnation, depending on your perspective—of effective public relations or risk communication. Morgan finds that when local residents talk about their community they code the plant positively, strictly because of its economic potential. Other potentials do not exist, or have been erased from the local discourse. I do not mean to suggest that Morgan's chapter is a paean to public relations so much as it is a statement that those who ignore public relations do so at their own peril. Although Morgan's demonstration that Carlsbad has become WIPP and WIPP has become Carlsbad is eerily disturbing, the more interesting (and politically relevant) aspect for me is his detailed description how they came to be so.

The differences between community discourse among residents around Carlsbad and discourse in the emerging downwinder community around Hanford is instructive. While the Carlsbad discourse aligns with a single code of economic advantage, the Hanford downwinders' discourse is never certain, always questioning, and explicitly stumbling over options for making sense of their mutual experience. Rather than assimilation, Kinsella and Mullen show us a simultaneous emergence of both organizational and individual identities. While their analysis joins other critical organizational analyses in encouraging us to reexamine notions of control and resistance, it never loses sight of the prevailing, respective positions of force (both material and symbolic). Members of the Hanford downwinder community do not fit into neat socioeconomic or political categories. They sometimes control and sometimes resist. Yet Kinsella and Mullen never let us forget that the DOE has been, and is, calling the shots.

Kinsella and Mullen's characterization of the downwinder community as emergent offers especially valuable insights for those studying organizational crisis management. Along with Taylor's, their analysis suggests that the nuclear weapons complex constitutes a *chronic crisis*, and identifies the need for an entirely new approach to studying the relationship between change and crisis. Their story details numerous layers of possibilities for change and how socioeconomic structures, political processes and communication practices come together to squelch its implementation. Although they do not retreat into a statement of unified dominance by the DOE and related interests, neither do they promulgate the attractive fiction of steady progress toward increased accountability, openness, and honesty among the leading organizations of the nuclear weapons complex. Instead, the crisis persists. Both Kinsella and Mullen's story of Hanford's downwinders and Taylor's critique of the DOE's Long-Term Stewardship discourse suggest the importance of examining the dangerous "fatigue, disillusionment, and burnout experienced by organizational change agents" (Taylor, et. al, 389). I can think of no more powerful analogue to the disturbing corporate takeover of civil society across the globe than the challenges of dealing with the "permanent emergency" generated by nuclear weapons production.

These critiques of the nuclear military complex suggests ways in which critical organizational scholars can move from the unfashionable (and admittedly simplistic) dominance model to an always dialectical sense of control and resistance (Mumby 2005), without jettisoning significant political engagement. They remind us that, although organizational authority and influence are fluid, during any given time and at any given place demarcations exist and sharply constrain possibilities for change and interaction among individuals, local communities, and nations.

NUCLEAR CRITICISM AND RHETORIC

Rhetoric traditionally has focused on communication as it is politically contested. Consistent with its connection to the classical tradition dating at least from the sophists' and Aristotle's attempts to systematize an approach that would enable citizens to participate rationally in their own governance, rhetoric's practical utility for citizens of democratic regimes has been its most enduring *raison d'être*. Despite the fact that rhetorical studies attempt to provide insights into contemporary political concerns, it is increasingly difficult for a growing number of citizens to imagine how these insights can be translated back to the practical, problem-oriented contexts where political deliberations transpire. Given its political roots and continued justification, I find the fact that rhetorical analysis sometimes offers limited textual insight, promising practical political value in some distant or idealized future, especially problematic.

The essays in this volume all reward the reader with an immediacy concerning problems confronting the variously embodied, situated, and interested audiences made up of citizens. Put otherwise, whereas appeals to practicality as a rationale for rhetorical studies often exhibit a highly imaginary quality, these examples of nuclear criticism provide a convincing demonstration of contemporary cultural significance—they matter here and now. The crux of my argument is that nuclear criticism can help to redeem the promise of practicality for rhetorical scholarship. Questions of the sort discussed in this volume are fundamental for practitioners of rhetorical analysis. They indicate how various approaches to political speech can contribute to a more lively public debate, as well as to its demise. They offer possibilities that could enrich the practice of rhetorical criticism at the same time they enliven the practice of political rhetoric.

Over the past few years, rhetorical scholars interested in political deliberation have generally come to rely on some variation of Habermas' public sphere to ground their analyses. That framework requires four basic conditions. First, all citizens must have meaningful opportunities to express their opinions to others. Second, citizens must deliberate rationally, temporarily setting aside differences in status. Third, they must be able to understand the issues being discussed. Finally, citizen participation must influence the development of public policy. Although the move to a framework for political reality based on Habermasian theory is an understandable response to a growing awareness that classical models fail to account for much of the technological rationality that permeates society, one does not have to be an expert in the rhetoric of science and technology to realize how unlikely it is for these conditions to be met. Further, we are guilty of the most callow nostalgia if we assume they ever have been met. Nuclear criticism projects such as the essays in this volume suggest a different approach to political relevance.

Taylor et al. (2005) argue that critical analysis of "U.S. nuclear weapons production . . . reveals a public sphere constricted and degraded by technocratic domination" (p. 281). The critical trajectory of these essays suggests how rhetoric might profitably respond to that technocratic domination while celebrating the fact that we can move the pieces around on the chessboard, and someone else may put us in checkmate tomorrow. Concepts such as stewardship, especially as examined by Taylor and McNamara, can programmatically justify the most inhumane acts, and we should be angry when this happens. But that anger can motivate us to the hard work needed to develop sufficient public expertise to challenge those acts. The Cold War triumphalism detailed by Krupar and Depoe can motivate strong counter-publics. The belief that science has betrayed its public interest can generate the rage needed to form the Hanford downwinder community.

Two examples strike me as immediately useful for rhetoricians seeking to enhance the practice of democracy in an age when democratic principles are publicly touted as endangering national security. First, Kinsella and Mullen provide a remarkably straightforward guide for anyone who seeks to encourage the development of a powerful counter-public. I say remarkably because they take great pains to avoid presenting Hanford downwinders as a monolithic force. Although they include numerous examples of internal contradictions among community members and detail complicating factors across time and space, they demonstrate how critics can use what could be seen as distractions, to strengthen the main story line. Their chapter exemplifies the political potential that justifies Ivie's (1996, 162) plea for an explicitly expanded notion of rhetoric as a constitutive cultural force "essential to the operation of the public sphere and the constitution of the polity."

The second example is Taylor's analysis of Long-Term Stewardship. Rhetoricians are fond of examining minute bits of language that resonate across cultures and issues, and the concept of stewardship is no exception. Humanities scholars produce tomes about god-terms such as community, liberty, and property. One might assume that, since these terms are so fundamental to U.S. political culture, the analyses would include practical discussions connected to the enhancement of such principles. Sadly, one's assumption would rarely be met. Taylor's essay demonstrates that critical analysis of such terms need not choose between theoretical sophistication, intellectual rigor, and immediate practicality. Readers are rewarded with a solid account of how stewardship has been framed, historical detail about how the DOE has reframed it, and evidence of the tensions that have emerged between stewardship and guardianship. All of this is linked back to democratic practice in a conclusion that argues that democracy's only hope lies in citizen responsibility to tirelessly interrogate the DOE's approach to stewardship.

Along with other essays in this volume, "Forever at Work in the Fields of the Bomb" suggests an approach to our constricted public sphere consistent with Burke's ironically hopeful "resignation to struggle" that recognizes democracy as a political process continually reinvented through cultural practices of implementation. The essays in this volume exemplify to varying degrees an interest in the intersection between communication and culture that increasingly characterizes contemporary rhetorical studies. They explore communicative practices for constituting human identity and political relationships. Their contemporary propinquity may have grown out of an intellectual shift that legitimizes communication's epistemic and ontological significance, a heightened awareness and distrust of science and technology, and/or an increasingly widespread discomfort with visible environmental degradation. Their focus on communicative exchange as social engagement among diverse participants who co-construct their situations offers a model that would be thoroughly consistent with the rhetorical tradition, and is fully cognizant of contemporary constraints. They suggest approaches that rhetorical scholars could take to help citizens better understand what public policy means in democracy. These stories have no conclusions because they do not end. Rhetorical scholars may discover in them inventional wisdom to strategically reconsider their political engagement in democratic practice.

NUCLEAR CRITICISM AS A CATALYST FOR COMMUNICATING DEMOCRACY?

The ideal democracy is not in hidden from our vision; it does not exist. Communication critics have a responsibility to remind us that every political system carries multiple entailments, and none are universally desirable. The best democracy can do for us is to surface hegemonic configurations that grow out of nepotism, corruption, and commercial pressures, and then offer space to renegotiate those configurations (Peterson et al. 2006). It cannot do away with them. If we accept a weakened notion of democracy as something fragile, to be protected from non-domesticated argument, it offers nothing. The essays in this volume suggest ways in which communication scholars can conduct research that is immediately relevant, especially in a time when dissent is discouraged, secrecy is demanded, and paranoia runs rampant. They must demonstrate that democracy without difference is irrelevant and tyranny is not synonymous with security.

Democracy is an adventure that both requires and restores "humility rather than arrogance, comedy rather than tragedy, and lunacy rather than rationality. In its full practice, we will collect bruises—not all of them metaphysical" (Peterson et al. 2006, 137). Nuclear criticism, environmental communication, organizational communication, and rhetoric offer diverse, and sometimes complementary, approaches to the always-present challenge of establishing democratic conditions that enable mutual recognition of self and other as participants in a political situation. Consciously synthetic application of these approaches may enable a robust approach to what Peters (1999, 230) claims is "the political question" of our age, that of "determining the range of creatures [with whom] we will communicate."

Beyond the obvious fact that we suffer individual harm when we lack air to breathe and water to drink, to the degree that we decimate the Earth we also decimate our own species. It does not matter whether we accomplish this task through nuclear weapons development or through environmental destruction and ambiguous remediation occasioned by that development. What does matter is our choice to engage the political process as self-conscious individuals and participants in a system structured largely by forces beyond our ken. Although the people behind those forces may operate from a completely altruistic desire to improve the lives of their fellows, we cannot afford to trust that to be the case. "The danger is that in the headlong rush toward progress, humans may destroy their grounding in nature under the delusion that natural and verbal systems are the same, and that they have the same freedom with, and power over, both" (Peterson 1997, 3). Against this danger democracy offers humility and uncertainty.

The essays in this volume demonstrate that, although the DOE possesses incredible resources of all kinds, and controls even more, it is not a monolithic entity. Its multiple and contradictory selves are revealed through analyses of communication practices that both grow out of and lead to the military industrial complex wherein we live. As with the residents of Carlsbad, it has become our place. These essays challenge us to commit ourselves to the unrelenting political activity of re-inventing our place.

BIBLIOGRAPHY

Ashcraft, Karen L. 2005. Resistance through consent? *Management Communication Quarterly* 19: 67–90.

Blair, Carole. 1999. Contemporary U.S. memorial sites as exemplars of rhetoric's materiality. Pp. 16–57 in *Rhetorical bodies*, ed. Jack Selzer and Sharon Crowley. Madison: University of Wisconsin Press.

Bryan, Todd A. and Julia M. Wondolleck. 2002. When irresolvable becomes resolvable: The Quincy Library Group conflict. Pp. 63–89 in *Making sense of intractable environmental conflicts: Concepts and cases*, ed. Roy J. Lewicki, Barbara Gray, and Michael Elliott. Washington, D.C.: Island Press.

Cox, J. Robert. 2007. Nature's "crisis disciplines": Does environmental communication have an ethical duty? *Environmental Communication: A Journal of Nature and Culture* 1: in press.

Depoe, Stephen P., ed. 2007. *Environmental Communication: A Journal of Nature and Culture* 1: in press.

Ivie, Robert L. 1996. Tragic fear and the rhetorical presidency: Combating evil in the Persian gulf. Pp. 153–178, 246–249 in *Beyond the rhetorical presidency*, ed. Martin J. Medhurst. College Station: Texas A&M University Press.

———. 2002. Rhetorical deliberation and democratic politics in the here and now. *Rhetoric and Public Affairs* 5: 277–285.

———. 2004. Prologue to democratic dissent in America. *Jasnost—The Public* 11: 19–36.

———. 2005. *Democracy and America's war on terror*. Tuscaloosa: University of Alabama Press.

Laclau, Ernesto and Chantal Mouffe. 2001. *Hegemony and socialist strategy: Towards democratic politics*. London: Verso.

Mouffe, Chantal. 1993. *The return of the political*. London: Verso.

——. 1996. Deconstruction, pragmatism and the politics of democracy. Pp. 1–12 in *Deconstruction and pragmatism*, ed. Chantal Mouffe. London: Routledge.

——. 2000. *The democratic paradox*. London: Verso.

Mumby, Dennis K. 2005. Theorizing resistance in organization studies. *Management Communication Quarterly* 19: 19–44.

Peters, John D. 1999. *Speaking into the air: A history of the idea of communication*. Chicago: University of Chicago Press.

Peterson, M. Nils, Markus J. Peterson, and Tarla Rai Peterson. 2005. Conservation and the myth of consensus. *Conservation Biology* 19: 762–767.

——. 2007. Environmental communication: Why this crisis discipline should facilitate environmental democracy. *Environmental Communication: A Journal of Nature and Culture* 1: in press.

Peterson, Tarla Rai. 1997. *Sharing the Earth: The rhetoric of sustainable development*. Columbia: University of South Carolina Press.

Peterson, Tarla Rai and Cristi Choat Horton. 1995. Rooted in the soil: How understanding the perspective of land-owners can enhance the management of environmental disputes. *Quarterly Journal of Speech* 81: 139–166.

Peterson, Tarla Rai, M. Nils Peterson, Markus J. Peterson, Stacy A. Allison and David C. Gore. 2006. To play the fool: Can environmental conservation and democracy survive social capital? *Communication and Critical/Cultural Studies* 3: 116–140.

Rogers, Richard A. 1998. Overcoming the objectifcation of nature in constitutive theories: toward a transhuman, materialist theory of communication. *Western Journal of Communication* 62: 244–272.

Taylor, Bryan C., William J. Kinsella, Stephen P. Depoe, and Maribeth S. Metzler. 2005. Nuclear Legacies: Communication, controversy, and the U. S. nuclear weapons production complex. Pp. 363–409 in *Communication Yearbook 29*, ed. P. Kalbfleisch. Mahwah, N.J.: Lawrence Erlbaum.

Tretheway, Angela. 2005. Reproducing and resisting the master narrative of decline: Midlife professional women's experiences of aging. *Management Communication Quarterly* 15: 183–226.

Williams, Terry Tempest. 2004. Ground truthing. *Orion*, 23/3 (May/June): 47.

Index

9/11, 24, 141, 189. *See also* terrorism; U.S. Department of Energy (DOE)

access to information, 46, 74
Adorno, Theodor, 5
anthropology: and communication scholarship, 170–71; and culture, 170; as critique, 171
Anti-Ballistic Missile Treaty, 13
argument spheres, 7. *See also* nuclear public sphere; public sphere
Atomic Energy Act, 58
Atomic Energy Commission. *See* U.S. Atomic Energy Commission
"atomic spaces," 111, 136, 140, 159
Atomic Testing Museum (ATM), 25 135, 136, 137, 141, 144, 159; autobiographical rhetoric, 96; design and exhibits, 143–44; funding sources, 142–43; "Ground Zero Theater," 135, 143, 144; as heritage preservation, 145; media coverage, 144; opposition, 144; origin, 141–42; proponents, 144–45. *See also* Hanford Downwinders

Bakhtin, Mikhail, 10
Battelle Laboratories, 83, 88, 98
Bechtel, Nevada, 141
Beck, Ulrich, 22, 79, 80
Berlin Wall, 140, 143
"boundary-work," 220–21
Burke, Kenneth, 4, 47, 76, 77, 79
Bush (G. W.): presidential administration, 1, 2, 62

Carlsbad (NM), 110, 122, 123, 124. *See also* Waste Isolation Pilot Plant (WIPP) site
Carlsbad Caverns National Park, 128–29
Chernobyl, 18, 89, 93, 98
China, 13
Citizens for Alternatives to Radioactive Dumping, 125–26
Clinton (W. J.): presidential administration, 58, 92, 180
Cold War, 1, 45, 75, 135, 136, 145, 169, 171, 178, 179, 200, 204, 205; conventions of secrecy, 75, 93, 94; culture, 18; end of, 11, 13, 16, 24, 25, 29, 140, 171; images of "veteran" and "hero," 223, 227n15; lessons of, 200; national security claims, 5–6, 82; nuclear arms race, 24; nuclear discourse, 29, 65;

studies, 140–41; "triumphalism" frame, 25, 140–41, 145, 158, 250; "victory," 221
collaborative learning. *See* public participation
Columbia River Advisory Group, 94
Columbia River United, 90
community groups near nuclear production sites, 46, 75
community-of-practice theory, 26, 174, 187; activity system, 177, 190; role of places 174; worldviews, 174
Comprehensive Environmental Response, Compensation, and Liability Act (CERCLA), 41, 50, 63; public participation activities, 50, 52. *See also* Fernald, Ohio site
Comprehensive Test Ban Treaty (CTBT) 12, 27, 169, 180–81, 184. *See also* nuclear weapons testing
consensus-building, 21. *See also* public dialogue; public participation
constitutive communication, 79, 99, 251
consubstantiality. *See* identification
containment, 6, 8, 45; discursive, 74, 80, 94, 139; of hazardous materials, 16, 74; rupture, 93, 95
corporate colonization, 6
"counter-monuments." *See* heritage preservation
counter-publics, 6, 250–51
critical rhetoric, 237
cultural anthropology, 26, 247
cultural discourse theory, 23, 111, 118–20; codes, 119, 121–24, 125, 128

deconstruction, 8
Deetz, Stanley, 6
deliberative process. *See* democratic practice; public deliberation; public participation
democratic practice, 237; mutual recognition of self and other, 252–53; perpetual change and tension, 239; role of conflict and dissent, 238–39, 242. *See also* public participation
Department of Energy. *See* U.S. Department of Energy (DOE)
Derrida, Jacques, 8, 10–11
Desert Research Institute (DRI), 142
deterrence policy, 65
dialogue, 6, 42, 43; consensus as goal, 65; exploration of differences, 44; normalized views of, 43; productive tension between convergence and divergence, 42, 44, 63, 64; recognition of dissensus, 44. *See also* democratic practice; public dialogue; public participation
dissemination model of communication, 241
Downwinders Opposed to Nuclear Testing (DONT), 144. *See also* Atomic Testing Museum; Hanford Downwinders

ecological communication, 22, 79
Einstein, Albert, 143
environment: danger of nuclear weapons, 252–53; material nature of ecological problems, 240
environmental communication, 28, 237; as crisis discipline, 245; environmental public sphere, 80; false dualism between humans and nature, 240–41; interplay between material and symbolic, 28, 240; mutual discursive influence among participants, 243; place and space, 243; stewardship trope in, 244–45; transhuman materialist rhetoric, 240
Espionage Act, 74
ethnography, 24, 25, 118, 120, 170
"evil empire," 1. *See also* Soviet Union

Federal Facility Compliance Agreement, 50, 58
Fernald, Ohio site, 14, 20, 21, 42, 46, 47, 64, 66, 87, 94, 95, 110, 114, 135,

136, 137, 242, 243; accelerated cleanup plan, 55–56; balanced approach, 56–57, 61, 63; CERCLA process, 50, 51, 52, 53, 54, 55, 59, 62, 63; cleanup budget cuts, 55; cleanup mission, 54, 146–47; closure, 150, 159n2; contextualized perspective, 52, 57; convergent views, 54, 57–58; de-contextualized perspective, 53; description, 145–46; differentiating past from present, 53–54; differentiating site from federal officials, 54–55, 61, 63; divergent views, 52, 59, 62; DOE Comprehensive Stewardship Plan, 60, 149; DOE Headquarters, 55, 56, 61; DOE Institutional Control Plan, 62; DOE Master Plan for Public Use, 61; DOE responsibility for contamination, 49; DOE site officials, 50, 53, 54, 59; environmental laws, 58; Feed Materials Production Center, 48; future use, 59–60; historic preservation activities, 147; legacy management institutional controls, 149; litigation, 49, 150; long-term stewardship, 59, 60, 61; media coverage of, 49; medical monitoring program, 49; multi-use education facility (MUEF), 25, 60, 136, 148–50; on-site disposal facility (OSDF), 50, 60; particularistic perspective, 62; post-9/11 shift in priorities, 62; production mission, 48, 146; public participation process, 50–51, 62–63; regulators' views, 50, 53, 55, 56, 59–60, 62, 67n4; restoration of aquifer, 50; scientific uncertainty, 53; scope and impact of environmental contamination, 48, 146–47; sense of place, 114–15; site-specific advisory board (SSAB), 51, 58–59 (*see also* Fernald Citizens Advisory Board); sustained stakeholder relationships, 63–64; waste disposal plan, 56

Fernald Citizens Advisory Board (FCAB), 147, 149, 150; "Future of Fernald" workshops, 148

Fernald Citizens Task Force (FCTF), 54, 56. *See also* Fernald Citizens Advisory Board (FCAB)

Fernald Living History (FLH), 66n2, 147, 149, 150

Fernald Living History Project (FLHP). *See* Fernald Living History (FLH)

Fernald Residents for Environmental Safety and Health (FRESH), 42, 47, 48, 146, 147, 150; focus on environmental and health issues, 58; formation, 49; forms of legitimation, 42, 47, 52, 57; participating inside and outside the process, 58; relationship with DOE, 52, 53, 54, 55, 56, 57, 59–60, 61, 64; views on future use, 60

Foucault, Michel, 9, 170–71, 178

frame of acceptance, 47

frames, 3, 4

Frank H. Rogers Science and Technology Building, 142

Freedom of Information Act, 82, 95

Goffman, Erving, 4

Goshute American Indians, 114

Groves, Leslie, 14, 73, 80, 82

Habermas, Jurgen, 5, 43, 250; communicative competence, 43; critique of, 44; ideal speech situation, 43

Hanford, Washington site 14, 46, 74, 75, 77, 80, 82, 84, 89, 94, 95, 98, 114, 128, 243, 248; Atomic Energy Commission cancer study, 88; Cold War legacy, 19; discursive rupture at, 94, 95, 97; environmental contamination and health hazards, 22, 75, 96; "Green Run"

experimental releases, 83, 89; heritage discourse, 99; information management at, 95; investigative journalism at, 82–83, 95; plutonium production, 75; public participation, 88; published accounts, 75; radiation exposures, 88; technical steering panel (TSP), 83, 88–89; Tri-Party Agreement, 98; whistleblowers, 80
Hanford Downwinders, 22, 74, 75, 76, 80, 81, 86, 87, 96, 144, 145, 243, 248, 250, 251; acknowledged as stakeholders, 91; association and dissociation, 79; Bailie, Tom, 82, 90, 95; Casey, June Stark, 85–86; crisis and change within, 99; development of a public voice and policy influence, 76; engagement in public nuclear discourse, 76, 90; Hanford Downwinder Coalition (HDC), 86, 92; "Hanford necklace," 84; Hanson, Gertie, 81, 82, 83, 92, 95; Hawkins, Ida, 90, 91, 95; health problems, 81, 84–85; identification with broader community, 76; initial awareness, 76; Jurji, Judith, 84, 91; litigation against DOE, 91–92; local knowledge, 90, 95; material consubstantiality, 77; narratives, 78–79, 93, 100; Pritikin, Tricia, 91–92; self-identification and organization, 75, 77; sense-making, 77; Smith, Millie, 86; stages of development, 75; stakeholder community, 93
Hanford Environmental Action League (HEAL), 82, 83, 84, 86, 94; public expertise, 95
Hanford Environmental Dose Reconstruction Project (HEDR), 22, 83, 89, 98, 101n8
Hanford Health Effects Subcommittee (HHES), 89, 91
Hanford Health Information Archives (HHIA), 89, 91
Hanford Health Information Network (HHIN), 89
Hanford Thyroid Disease Study (HTDS), 22, 83, 90, 98, 102n8
hegemony, 8, 9–10
heritage preservation, 25, 138; "counter-monuments," 138; preservation activities, 138, 139–40; strategies of remembrance, 138
Hiroshima, 8, 17, 73, 74, 89, 139
history, 137; critical, 137; ideological, 137, 141, 158; rhetorical uses, 141
Horkheimer, Max, 5

identification, 76–77, 100; through antithesis, 46; and consubstantality, 77; identity construction, 247
Institute for Energy and Environmental Research (IEER), 95
institutional communication, 6
instrumental rationality, 6
Iranian government, 1
isolation, 109, 110

Kaiser-Hill Company, 153, 154
knowledge: artifacts vs. texts, 187–88; as a commodity, 26, 187; as a discursive regime of truth, 9, 169; and identity, 173–74; local, 80, 90; and power, 171, 178; preservation, 26, 182, 185; tacit, 168, 174; traditional view of science, 168; ways of knowing, 26, 169, 193
Kuhn, Thomas, 172

Lawrence Livermore National Laboratory (CA), 45, 113, 143, 167, 175, 190, 191, 192, 212; Nuclear Weapons Information Group (NWIG), 183; Nuclear Weapons Information Project (NWIP), 183
legitimacy, 8
legitimation crisis, 97, 99
Lockheed Martin Nevada Technologies, 142

Long-Term stewardship (LTS), 27, 28, 59, 60, 61, 66, 97, 200, 201, 206, 210, 213, 217, 218, 225, 249, 251; accelerated cleanup and, 208, 216; burden minimization, 215–16; challenges to optimizing LTS, 211–13; DOE scoping workshops, 209; domesticated discourse, 214–15; funding lifecycle, 208, 218–19; ideals for optimizing LTS, 210–11; levels of stewardship, 207; opportunities for stakeholder participation, 216–17, 224–25; origin within DOE, 207; relationship to cleanup and end-states, 213–14; stakeholder discourse, 208, 209; time and, 217–19. *See also* stewardship; Stockpile Stewardship Program (SSP); U.S. Department of Energy (DOE)
Los Alamos National Laboratory (LANL), 14, 25, 113, 167, 171, 173, 175, 176, 177, 178, 179, 180, 182, 184, 185, 188, 189, 191, 193, 212; Nuclear Weapons Archiving Program (NWAP), 183; security problems, 184; Theoretical Institute for Thermonuclear and Nuclear Studies (TITANS), 183, 190, 191, 192
Luhmann, Niklas, 22, 79, 80, 96

Manhattan Project, 15, 18, 73, 129
memory, 137; collective, 137–38, 139–40, 159; forgetting, 138; public, 138

Nagasaki, 8, 18, 73, 74, 75, 139
naming, 4
narrative paradigm, 78–79; narrativity and, 78–79, 100
narratives, 5, 78–79, 93, 100
National Academy of Sciences, 98, 122
National Association of Radiation Survivors, 90
National Defense Authorization Act of 1991, 91
National Lead of Ohio, 48. *See also* Fernald, Ohio site
National Nuclear Security Administration (NNSA), 17, 142, 192, 212, 237
National Priorities List, 50, 58
national security, 73, 169, 189, 193, 200; post-Cold War, 180
Nevada Test Site (NTS), 82, 89, 135, 141, 145, 167, 173, 176, 178–79, 185–86; moratorium, 141, 142; testing function, 141
Nevada Test Site Historical Foundation (NTSHF), 142
North Korea, 1, 13
Northwest Radiation Health Alliance, 96
nuclear communication scholarship, 9, 11, 44. *See also* nuclear criticism; nuclear weapons and communication
nuclear confidence, 167, 168, 169, 175, 185, 188; development of new activity system, 188; modeling uncertainty and, 191, 192; threats to, 181
nuclear criticism, 4, 9, 28, 245, 250; argumentation theory and analysis, 6–7; dramatistic tradition, 4–5; modernist critical theory, 5–6; semiotic, post-structuralist, and dialogic theories, 7–11
nuclear deterrence, 137, 167, 169, 171, 175, 178, 191, 200, 205; credibility, 169, 175, 176
nuclear discourse, 43, 47
nuclear elites, 26
nuclear energy, 73
Nuclear Freeze, 6
nuclear fuel cycle, 15
nuclear guardianship, 204–6, 223–24, 225, 226n6
nuclear legacies, 2–3, 8, 17–19, 219
Nuclear Non-Proliferation Treaty (NPT), 12, 31n13
nuclear policy discourse, 7
nuclear proliferation, 1, 2, 13, 189

nuclear public sphere, 74, 76, 80
nuclear themes or tropes: entelechy, mystery, potency, secrecy, 2, 77, 126–27, 139; time and space, 19
nuclear waste, 16, 18, 114, 217; transuranic, 16, 23
nuclear waste repositories. *See* Waste Isolation Pilot Plant (WIPP); Yucca Mountain site (NV).
nuclear weapons, 4; arsenal, 136; civilian and military systems of control, 205, 206; future development, 26; "perfection" of technology, 4; possessing, making, and cleaning up, 199; program to disassemble, 15; situated in time, 206; stockpile, 167, 188, 191; symbol of dangerous technology, 204; systems, 169; terrorists and, 15; use, 8–9; in U.S. foreign and military policy, 199; weapons-grade materials, 15
nuclear weapons and communication, 138–40; dialogic focus, 139; domestication, 129n2; "fabulously textual," 8; forgetting, 129; groups unable to reach common understandings, 45; material and symbolic dimensions, 138–39; performance, 113; site and place, 128; struggle among competing interpretations, 45; technostrategic communication, 113–14. *See also* nuclear criticism
nuclear weapons history/museums, 136
nuclear weapons laboratories: anthropological studies of, 171; design-and-test cycle, 176–77, 187; impact of knowledge loss, 168, 181; knowledge preservation efforts, 182–86; regime of truth, 178, 193; science as source of credibility, 176; secrecy within, 189–90. *See also* Lawrence Livermore National Laboratory (CA); Los Alamos National Laboratory (LANL); Sandia National Laboratory (NM)
nuclear weapons production complex, 2, 3, 10, 11, 20, 48, 93, 111–12, 135, 159, 167, 175, 237, 239, 243, 246, 248; as chronic crisis, 248, 249; commercial production sites, 112–13; environmental remediation sites, 114; historical period of operations, 24; impact of production on public health, environment, and worker safety, 8, 15, 136, 200, 206; lack of environmental oversight, 48; military production sites, 113–14; places of nuclear maintenance, 114; production and remediation missions, 199; risks and benefits of production for communities, 221–22; sites of multiple and conflicting voices, 10; as threatening arsenal and industrial and cultural life-world, 11; workers, 110, 136
nuclear weapons production complex sites. *See* Fernald, Ohio site; Hanford, Washington site; Lawrence Livermore National Laboratory; Los Alamos National Laboratory; Oak Ridge, Tennessee site; Pantex, Texas facility; Rocky Flats, Colorado site; Sandia National Laboratory; Savannah River, South Carolina site
nuclear weapons professionals, 9–10
nuclear weapons testing, 1, 25–26, 135, 167; ban, 169, 179; Hatfield Amendment, 167, 179; moratorium, 167; reliability and performance, 168, 178; source of confidence, 176, 178–79. *See also* Comprehensive Test Ban Treaty (CTBT)
"nukespeak," 7, 139, 206. *See also* nuclear criticism; nuclear weapons and communication

Oak Ridge, Tennessee site, 94, 113
Ohio Environmental Protection Agency, 42, 47, 50, 53, 55, 56, 149. *See also* Fernald, Ohio site
O'Leary, Hazel, 58, 97, 203
openness, 41. *See also under* U.S. Department of Energy
organizational communication, 245; critical studies of control and resistance, 245–46

Pacific Northwest National Laboratories (PNNL), 83, 98
Pantex, Texas facility, 15, 113
Perelman, Chaim, 6
Physicians for Social Responsibility, 88, 95–96
popular epidemiology, 81
post-Cold War era, 2, 3, 9, 46, 76, 199; changing missions and budget, 12; decline in global nuclear arsenal, 12; development of smaller-yield nuclear weapons, 13; failure to negotiate arms-reduction treaties, 13; inertia of nuclear-military bureaucracies, 13; neo-conservative narratives, 13; nuclear weapons discourse in, 206; nuclear weapons production, 6, 199; securing nuclear stockpiles, 12, 13–14; threat of "dirty bomb," 13; threat of horizontal nuclear proliferation, 13
power, 6, 10, 20
public deliberation, 5, 7. *See also* public participation
public dialogue, 20; consensus-building and conflict, 21, 22. *See also* dialogue; public participation
public expertise, 22, 81, 95, 100; challenges technocratic norms, 96; through local knowledge, 95; through participatory research, 96; through technical assistance, 95
public participation, 27, 41, 42, 216–17, 224–25, 237; collaborative learning, 43; consensus building approach, 42; democratic theories of, 43; dialogue with divergence, 63; evolving pattern of convergence and divergence, 63; fair and competent process, 43; long-term interaction, 63; measuring effectiveness, 46, 65; mechanisms, 42–43; one-way versus two-way, 43; trinity of voice, 81. *See also* democratic practice; dialogue; public deliberation
public sphere, 5, 250

Quantifying Margins of Uncertainty (QMU), 192
Quincy Library Group (QLG), 244

Radiation-Exposed Veterans Compensation Act, 91
radiation hazards, 74
Radiation Victims Fair Treatment Act, 87
rational-world paradigm, 78, 96
Reagan (R. W.): presidential administration, 65, 146
Reliable Replacement Warhead (RRW), 169, 192. *See also* nuclear weapons
rhetoric, 249, 251
risk community or "risk society," 22, 76, 79, 80, 100; reflexive modernity, 79
Rocky Flats, Colorado site 14, 19, 87, 94, 98, 114, 115, 135, 136, 137, 214; closure, 159n2; competing visions of site's future, 152; DOE museum feasibility report, 155; environmental and health risks, 152; FBI raid, 152; museum/education center, 136, 153; National Wildlife Refuge Act, 153, 154; plutonium fires of 1957 and 1969, 151; production mission, 151; remediation efforts, 152–53
Rocky Flats Citizens Advisory Board (RFCAB), 152, 154
Rocky Flats Cold War Museum (RFCWM), 25, 154, 155; artifacts

and exhibits, 155–56; DOE resistance, 155; funding, 154, 155; land, 156; mission, 154; museum feasibility study, 154; political support, 154
Rocky Flats History Project. *See* Rocky Flats Cold War Museum (RFCWM)
Rocky Flats Virtual Exhibit or Museum, 25, 156–58; as counter-statement, 158; depiction of 1969 fire, 157–58; origin, 156–57
Roosevelt, Franklin, 14, 113, 143
Russia, 12
Ruttenbur, James, 84, 85

Sandia National Laboratory (NM), 113, 175, 183, 184, 191, 193n4, 212; Knowledge Preservation Project (KPP), 183
Savannah River, South Carolina site, 14, 113
Science-Based Stockpile Stewardship (SBSS), 167, 168, 180, 185. *See also* Stockpile Stewardship Program (SSP)
scientific knowledge: constitution of communities, 172; sociology of, 172, 174; *See also* knowledge
scope of identification, 47, 66
secrecy, 6, 41. *See also* Cold War; U.S. Department of Energy (DOE)
self-organization, 74
sense-making, 22, 77, 100, 117
sense of place, 23, 110, 114–15
site-specific advisory boards (SSABs), 16, 46, 58, 209; effectiveness, 46; evaluation research, 46
Smithsonian Institution, 139
Smyth Report, 73–74, 83
Soviet Union, 12, 25, 99, 136, 140, 141, 189, 190
stakeholder perspectives, 42
stakeholders and stakeholder participation, 6, 16, 20, 21, 42, 43, 47, 207, 210, 216–17, 224–25, 242, 244. *See also* democratic practice; dialogue; public participation
stewardship, 27, 200; etymology and traditional usage, 201, 219; future-oriented, 204; mythical narratives in, 201–2, 215; polysemic and dialogic meanings, 201, 205–6, 220; site of struggle, 202, 219; tension with guardianship, 205, 220, 225. *See also* Long-Term Stewardship (LTS); Stockpile Stewardship Program (SSP)
Stockpile Stewardship Program (SSP), 180–81, 184, 191, 203, 222, 224, 225; competing with LTS, 222–23; environmental risks, 222; importance of maintaining institutional knowledge, 184; need for stakeholder participation, 224–25; redefining missions, 184, 185–86, 190
Strategic Arms Reduction Treaty (START), 141
Superfund. *See* National Priorities List
symbolic convergence theory, 79

technical and cultural rationality, 46, 80
technical sphere. *See* argument spheres
Teller, Edward, 17
terrorism, 29, 212–13. *See also* 9/11; U.S. Department of Energy (DOE)
Three Mile Island, 82, 83, 89
Till, John, 88
Titus, Dina, 143
Toulmin, Stephen, 6
Trinity atomic test, 75

U.S. Agency for Toxic Substances and Disease Registry (ATSDR), 89
U.S. Army Corps of Engineers, 145
U.S. Atomic Energy Commission (AEC), 15, 48, 88, 94, 122, 136, 15
U.S. Centers for Disease Control (CDC), 83, 84, 90, 98
U.S. Congress Office of Technology Assessment (USOTA), 19

U.S. Department of Defense (DOD), 116, 146
U.S. Department of Energy (DOE), 16, 21, 41, 45, 46, 48, 49, 58, 75, 77, 78, 79, 81, 83, 87, 88, 91, 92, 95, 97, 98, 110, 113, 116, 135, 152, 158, 159, 167, 176, 180, 200, 201, 206, 208, 210, 211, 212, 213, 216, 217, 218, 222, 237, 242, 251, 253; "closed circle" of weapons cleanup, 16; Cold-War era organizational culture, 212, 221; community relationships, 42; environmental cleanup decisions, 41–42; environmental remediation mission, 136; expand mission from deterrence to environmental protection, 200; Headquarters, 42, 55; Headquarters vs. site officials, 21; heritage preservation activities, 159; impact of 9/11 terrorist attacks on openness, 212–13; Internet database of waste holdings, 208–9; Long-Term Stewardship Study, 216; medical experiments, 87; Office of Defense Programs (ODP), 182, 222, 223; Office of Environmental Management (EM), 67n5, 149, 207, 215, 222, 227n10; Office of Legacy Management (LM), 155, 156, 208; Office of Long-Term Stewardship, 208; "openness" initiative, 16, 58, 97, 203, 223; organizational crisis, 97; post-Cold War efforts, 17; public distrust and skepticism, 42; public participation programs, 46, 65, 67; site officials, 47; stewardship program (*see* Long-Term Stewardship [LTS] and Stockpile Stewardship Program [SSP]); Stockpile Lifetime Extension Program (SLEP), 182; technocratic logic, 78; Environmental Protection Agency (EPA), 16, 41, 47, 109, 127, 147. *See also* National Nuclear Security Administration (NNSA), U.S.
U.S. Forest Service, 244
U.S. Nuclear Regulatory Commission (NRC), 112, 113, 207
U.S. Public Health Service, 94

vernacular discourse, 6

Waste Isolation Pilot Plant (WIPP) site, 16, 23, 109, 110, 111, 120, 121, 122, 126, 127, 243, 248; code of economy in discourse, 23, 121–24, 125, 128; code of nothingness in discourse, 23, 124–25, 128; Environmental Evaluation Group, 109; isolation, 23; location, 116; opposition, 125–26; site characterization, 116–17; as structure of everyday life, 116, 129; transuranic waste, 110, 117
World Trade Towers, 143
World War II, 138, 140, 221

Yucca Mountain (NV) site, 16, 114

About the Contributors

Stephen P. Depoe (Ph.D., Northwestern University, 1986) is associate professor in the Department of Communication at the University of Cincinnati, where he also directs the Center for Health and Environmental Communication Research. He is a cofounder and the current editor of *The Environmental Communication Yearbook*, and founder of the new journal *Environmental Communication: A Journal of Nature and Culture*. Depoe's published work in environmental communication includes the edited volume *Communication and Public Participation in Environmental Decision Making* (2004), along with other publications related to history and heritage preservation within the nuclear weapons complex. He has served as a member of the Fernald Citizens Advisory Board since 2000, and was President of Fernald Living History Inc. from 2001 to 2005.

Jennifer Duffield Hamilton (Ph.D., University of Cincinnati, 2003) is a research fellow at the Center for Health and Environmental Communication Research at the University of Cincinnati. Her research explores public participation and risk communication in environmental decision making, especially within the nuclear weapons complex, and the theory and practice of dialogue in environmental policy contexts. Her work has appeared in *Risk Analysis*, *Environmental Communication Yearbook*, and *Communication and Public Participation in Environmental Decision Making* (2004). She participated in developing the Fernald Living History Project at the Department of Energy's Fernald site in Ohio and has followed cleanup issues there for a number of years.

William J. Kinsella (Ph.D., Rutgers University, 1997) is a faculty member in the Department of Communication and the interdisciplinary program in Science,

Technology and Society at North Carolina State University. His research examines the intersections among organizational communication, environmental communication, rhetoric of science and technology, and rhetoric of public policy. Kinsella earned a B.S. degree in physics, did graduate studies in astronomy and physics, and worked as a science educator before transitioning into the communication field. His publications have appeared in *Communication Yearbook, Environmental Communication Yearbook, Management Communication Quarterly*, *New Jersey Journal of Communication, Qualitative Research Reports in Communication, Rhetoric Society Quarterly*, *Science as Culture*, *Technical Communication Quarterly,* and two academic books. From 2000 to 2006 he served on the citizen advisory board for the Department of Energy's Hanford site.

Jason N. Krupar (Ph.D., Case Western Reserve University, 2000) is assistant professor of history at the OMI College of Applied Science, University of Cincinnati. His current research interests include historic preservation efforts within the nuclear weapons complex, Cold War science and technology policies, urban planning and technology, and atomic culture issues. His recent projects include examining the leadership and organizational culture of the U.S. Atomic Energy Commission (AEC) and race relations/policies within the Manhattan Project and early AEC, circa 1950s. He has served on the Board of Directors of the Rocky Flats Cold War Museum.

Laura A. McNamara (Ph.D., University of New Mexico, 2001) is a principal member of technical staff at Sandia National Laboratories in Albuquerque, New Mexico. An applied anthropologist, her research interests include expert knowledge elicitation, knowledge representation, distributed cognition, collaboration, uncertainty quantification, and verification and validation for social and behavioral modeling and simulation technologies. McNamara has done fieldwork in the U.S. Missile Defense Agency, the Defense Intelligence Agency, and the nuclear weapons programs at Sandia and Los Alamos National Laboratories. Her research has been published in *Reliability Engineering and Systems Safety*, *Technometrics*, and the *Engineering Design Reliability Handbook*. She also edits a book series on interdisciplinarity for Elsevier Press, which recently released her coedited volume (with Ray Paton) entitled *Multidisciplinary Approaches to Theory in Medicine*.

Maribeth S. Metzler (Ph.D., Rensselaer Polytechnic Institute, 1996) is associate professor and director of the public relations program at the University of Cincinnati. Prior to her doctoral studies, she worked in environmental public affairs for the U.S. Air Force and as an environmental consultant. Her research focuses on the social implications of organizations and has appeared in *American Behavioral Scientist*, *Communication Quarterly*, *Communication Studies*, *Communication Yearbook*, *The Encyclopedia of Public Relations* (2004), *The Handbook of Public Relations* (2001), and *Responsible Communication: Ethical Issues in Business, Industry, and the Professions* (1996).

Eric L. Morgan (Ph.D., University of Massachusetts, Amherst, 2002) is assistant professor in the Department of Communication Studies at New Mexico State University. His research program is primarily focused on the cultural construction of place as investigated through the ethnography of communication. His work includes studies of sense of place in western Massachusetts, southern New Mexico, and Arctic Canada.

Jay Mullen (Ph.D., University of Kentucky) is professor of history at Southern Oregon University. He received a B.A. from the University of Oregon, and received an M.A. and a Ph.D. in history from the University of Kentucky. Mullen has been a National Defense Foreign Language Fellow in Wolof at Indiana University's African Language Institute, a Fulbright scholar in Egypt, and a National Endowment for the Humanities Fellow at the University of Texas.

Tarla Rai Peterson (Ph.D., Washington State University, 1987) holds the Boone and Crockett Chair of Wildlife and Conservation Policy at Texas A&M University. She also is an adjunct professor in the Communication Department at the University of Utah. She received an Interdisciplinary Ph.D. Her teaching and research interests revolve around expanding opportunities for people to participate in the development, implementation, and evaluation of public policy designed to provide an environment where people and other creatures can thrive. Rai Peterson has published the results of her research in *Sharing the Earth: The Rhetoric of Sustainable Development* (University of South Carolina Press, 1997), as well as in scholarly journals including *Agriculture and Human Values*, *Communication and Critical/Cultural Studies*, *Conservation Biology*, *Ecological Modeling*, *Environmental Practice*, *Environmental Values*, *The Quarterly Journal of Speech*, *Journal of Applied Communication Research*, *Journal of Wildlife Management*, as well as in several book chapters and symposium proceedings.

Bryan C. Taylor (Ph.D., University of Utah, 1991) is associate professor of communication at the University of Colorado, Boulder. His primary research interests include cultural memory of the Cold War and organizational cultures associated with nuclear weapons development. His work on these topics has been published in *American Literary History*, *Critical Studies in Media Communication*, *Journal of Contemporary Ethnography*, *Journal of Organizational Change Management*, *Quarterly Journal of Speech*, *Rhetoric and Public Affairs*, *Studies in Cultures, Organizations and Societies*, *Western Journal of Communication*, and elsewhere. He is coauthor with Thomas Lindlof of *Qualitative Communication Research Methods*, 2d ed. (2002).